Proceedings of the Munich Symposium on Lightweight Design 2024

Felix Endress · Jasper Rieser · Klemens Rother ·
Philipp Höfer · Tobias Dickhut ·
Markus Zimmermann
Editors

Proceedings of the Munich Symposium on Lightweight Design 2024

Springer

Editors
Felix Endress
Technical University of Munich
Garching, Germany

Klemens Rother
Munich University of Applied Sciences
Munich, Germany

Tobias Dickhut
Ludwig Bölkow Campus
Universität der Bundeswehr München
Taufkirchen, Bayern, Germany

Jasper Rieser
Technical University of Munich
Garching, Germany

Philipp Höfer
Universität der Bundeswehr München
Neubiberg, Bayern, Germany

Markus Zimmermann
Technical University of Munich
Garching, Germany

ISBN 978-3-032-01359-0 ISBN 978-3-032-01360-6 (eBook)
https://doi.org/10.1007/978-3-032-01360-6

Responsible Editor: Alexander Grün

This Springer imprint is published by the registered company Springer Nature Switzerland AG
The registered company address is: Gewerbestrasse 11, 6330 Cham, Switzerland

Preface

Dear reader

The Munich Symposium of Lightweight Design (former "Leichtbauseminar") for more than two decades is an annual conference on lightweight design hosted by Technical University Munich, Universität der Bundeswehr München and Munich University of Applied Sciences, all three located at Munich.

The Symposium focusses on the reduction of energy consumption and increase of product performance by reducing mass through lightweighting and thus also motivates resource savings for mechanical structures of all kinds of industrial branches. Researchers, students and members from industry mutually join this conference for exchange and discussion on such topics.

In the present volume you will find many of the presentations at the twenty-first conference of this series given on three exciting afternoons. Special thanks to the team from Springer for supporting this publication and for giving excellent support as well as Jasper Rieser and Felix Endress for conducting the review processes and the organization of the seminars and the proceedings.

The 21st symposium again features many aspects on lightweight design like strength and manufacturing of components made of fiber composites, weight reduction of electric drives, strength improvement of titanium alloys, control-co-design for wing structures, application and analysis of sandwich design, numerical optimization, analysis and testing of added manufacturing or conceptual development of giga-castings. Not all presentations are contained in these proceedings though.

As a keynote for the 2024 symposium, Prof. Dr.-Ing. habil. Ralf Cuntze, a respected engineer and pioneer for composite design and analysis, presented a *Comparison of four UD Strength Criteria—including some Review*. You will find this presentation—among many more publications of Prof. Cuntze under the following link: https://www.carbon-connected.de/Group/Prof.Ralf.Cuntze/Pages/Start/Index/174916.

Prof. Dr.-Ing. habil. Ralf Cuntze participated and supported the Munich Lightweight Symposium from the very beginning as a participant as well as presenter many times. His contributions to discussions were always a great asset to these events and will be remembered by many. His two outstanding presentations at the 20th and 21st symposium should be seen as a special appreciation of Ralf Cuntze's work for lightweight construction in general and fiber composite technology in particular.

As a wish by all editors, you hopefully will find some ideas and inspiration for developing better, lighter and more performant structures as one building block for saving earth's climate and resources.

Klemens Rother
On behalf of all Editors of the Munich
Lightweight Symposium

Contents

Path Planning of Load-Path-Adapted Variable Angle Tows in Automated Fiber Placement

Raphael Höfer[1](✉), Pezhman Pourabdollah[1], Felix Gehlhoff[1], and Alexander Fay[2]

[1] Helmut-Schmidt-University, Holstenhofweg 85, 22043 Hamburg, Germany
raphael.hoefer@hsu-hh.de

[2] Ruhr-Universität Bochum, Universitätsstraße 150, 44801 Bochum, Germany
https://www.hsu-hh.de/aut/

Abstract. Fiber-reinforced composites are crucial in many different industries due to their high strength-to-weight ratio. Automated Fiber Placement (AFP) enables precise deposition of carbon fiber tapes to create laminates, yet fixed fiber orientations often misalign with complex aerospace load paths, leading to suboptimal load transfer and higher weight of the component. Aligning fibers with actual load paths enhances performance by improving load transmission thus reducing thickness of the laminate. This paper introduces a path planning method that generates curvilinear fiber paths using Finite Element Analysis (FEA) results. These paths align with load directions while adhering to AFP manufacturing constraints, ensuring a defect-free production.

Keywords: Composite laminate · Variable stiffness · Curvilinear fiber · Manufacturing constraints · Automated Fiber Placement · Path planning

1 Introduction

The utilization of components produced with Carbon fiber-reinforced plastics (CFRP) has become a crucial part in the domain of lightweight design, particularly in sectors where weight is a critical factor, such as aviation and aerospace [1–3]. Due to manufacturing limitations and domain-specific regulations these components are typically produced using unidirectional plies consisting of standard orientations of 0°, 90° and ±45°, resulting in quasi-isotropic laminates [4].

While quasi-isotropic laminates distribute the loads evenly along the component, they do not fully take advantage of the unique properties of CFRP materials to transmit forces along the fiber direction. Because of this it is beneficial to align the fibers with the load path. This limitation is particularly noticeable in aerospace applications, where load paths can be complex and vary significantly between components [4]. When there is a misalignment between the material's

F. Endress et al. (Eds.): MSLD 2024, *Proceedings of the Munich Symposium on Lightweight Design 2024*, pp. 1–8, 2025.
https://doi.org/10.1007/978-3-032-01360-6_1

deposition direction and the actual load path, it often requires thicker and heavier laminates in order to withstand the forces because of inefficient load transfer, which can undermine some of the weight-saving benefits of CFRP.

Advanced manufacturing techniques like Automated Fiber Placement (AFP) provide a solution to this issue by allowing the deposition of carbon fiber tapes in customized, directed orientations. AFP, as shown in Fig. 1a is a process where pre-impregnated carbon fiber tapes are automatically placed onto a mold or tool in successive layers, guided by robotic systems [5,6]. This technique enables precise control over fiber placement, reducing material waste and improving structural performance. By adjusting the fiber paths to closely match the actual load directions, these techniques enhance load transmission, reduce laminate thickness, and create lighter components [7]. However, this method also brings new challenges in path planning, as the creation of curvilinear paths must comply with strict AFP manufacturing constraints to prevent defects such as gaps and wrinkles, due to introducing compressive forces on the inner and tensile forces on the outer side of the deposited tape [8], as shown in Fig. 1b. These defects lead to a reduction in the laminate's structural performance and should therefore be avoided [9].

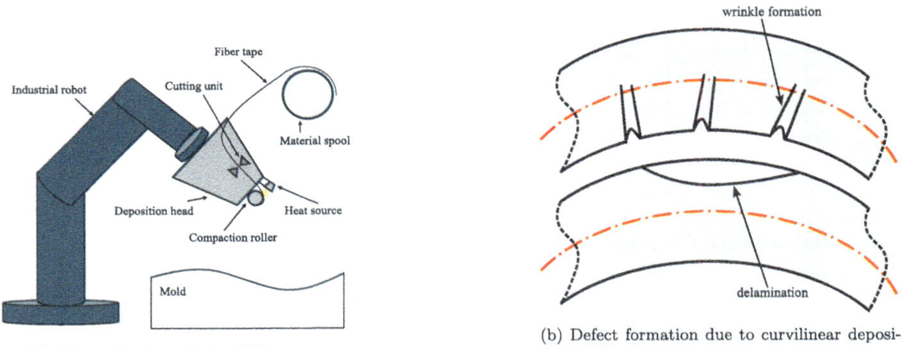

(a) Schematic of a robotic AFP process.

(b) Defect formation due to curvilinear deposition of fiber tapes.

Fig. 1. Illustrations related to the AFP process.

Coupling FEA-derived load data with curvilinear path planning methods will provide fiber paths that more closely follow the structural requirements of the components while maintaining manufacturability. This represents a major step toward taking full advantage of the unique properties of CFRP in high-performance and lightweight design applications.

In this paper, a new directionally constrained path planning approach is presented, utilizing results from FEA in order to generate load-path-adapted fiber paths for AFP. The remainder of the paper is structured as follows: Sect. 2 describes the current state of the art in curvilinear path planning, followed by Sect. 3, where the path planning algorithm is introduced. Section 4 gives a brief summary of the presented work.

2 Variable Angle Path Planning in Automated Fiber Placement

Path planning in AFP plays a crucial role in ensuring that CFRP meet both structural and manufacturing requirements. The process involves determining the precise trajectories the robotic system has to carry out for depositing the carbon fiber tapes along the generated fiber paths, which at best align with the load paths of the component while adhering to manufacturing constraints such as minimum steering radius, maximum allowable gaps, and overlap tolerances. Effective path planning directly influences the mechanical performance, weight efficiency, and defect-free production of the laminate.

Traditional approaches to path planning often rely on predefined fiber orientations, which can result in suboptimal alignment with complex load paths. This limitation has driven the development of advanced computational methods that leverage optimization algorithms and FEA to generate steered fiber trajectories. These methods aim to align fiber paths with principal stress directions, thereby enhancing load transfer efficiency.

This section delves into recent studies of AFP path planning, explores state-of-the-art methodologies, and highlights the importance of integrating structural analysis with manufacturing constraints in the path planning. By addressing these challenges, researchers and engineers can fully exploit the potential of CFRP materials for high-performance applications.

In [10] the authors clustered the considered component into different segments using a normalized cut segmentation algorithm based on the load path extracted from FEA. Then, for each segment, a reference curve is generated starting from the focal point of each segment following the load path by fitting a curve to the load path vector field. These reference curves are then populated over each segments. The authors accomplished an effective curvilinear path planning considering the minimal layup length of the used system as well as minimum steering radii resulting in a defect free variable angle ply without defects. But further development needs to be carried out to verify the mechanical properties of the manufactured part as well as to transfer the presented method on curved parts.

In [7] the authors further developed the method mentioned above by performing a lamination parameter optimization before segmentation and path generation. Based on the optimization, discrete fiber angles are derived and further path planning is carried out. The authors achieved an improvement of up to 14.3% on a conical shape compared to a unidirectional laminate design. But still the validation if the mechanical properties have to be verified in practice.

Shahabad et al. [11] present two state-of-the-art methods that use lamination parameters to extend the design space for variable stiffness laminates, namely the Lamination Parameter Extrapolation Method (LPEM) and the Relaxed Lamination Parameter Interpolation Method (RLPIM). While LPEM extends the optimization domain beyond pre-defined bounds, RLPIM allows parameters to vary with different volumetric ratios. Both use the spectral Chebyshev approach to achieve high computational efficiency, allowing reductions in simulation time while ensuring continuity constraints and thus manufacturable fiber paths. However, no experimental or manufacturing applications have yet been performed in order to validate it practically, which can be a promising direction for further research.

A level set of a real-valued function is where a function takes on a given constant value [12]. Level set methods have recently emerged as effective tools for representing curvilinear fiber paths, providing a unified framework for optimization without requiring complex post-processing. However, earlier implementations often lacked precise handling of curvature constraints, which could compromise manufacturability.

The work of Tian et al. [13] extends the application of level set methods by incorporating compactly supported radial basis functions and taking the iso-contours of those functions for accurate fiber path representation and optimization. The iso-contours insure a gap-free path generation and by explicitly enforcing gap, overlap and curvature constraints, the method ensures manufacturable fiber paths while achieving significant performance improvements. However, the approach has yet to be validated with practical manufacturing and experimental results, leaving an important area for future exploration.

Another method using level set approach is presented by Arsenyeva [14]. Here, the iso-contours for curvilinear path generation are derived by optimizing the control points of an artificial smooth surface for optimal fiber direction while considering minimal steering radii. The authors validated the presented method by applying the approach to a flat and curved component. Results show an increase in the components performance. Further comparison to existing techniques as well as practical layup has to be carried out in the future.

3 Load-Path-Adapted Path Planning for Defect-Free Deposition

A significant portion of variable angle path planning literature focuses on optimizing fiber paths, a process demanding extensive computational resources. This demand limits scalability, hinders real-time applications, and increases development costs and time for efficient fiber placement strategies. The time required depends on the number of optimization variables, with higher complexity directly increasing computational needs. While optimizing lamination parameters can reduce design variables, it still requires substantial effort [15]. To address this, a novel path planning method is essential to minimize time and effort. The next chapter introduces an approach for load-path-adapted planning for defect-free deposition in AFP, incorporating geometric and directional constraints.

3.1 Quantifying Curvature to Address Manufacturing Constraints

For a defect-free deposition of the material we need to consider certain manufacturing constraints. In this case the steering radius of the robots trajectory is the critical parameter for defect formation during the layup. A smaller steering radius leads to a higher occurrence of gaps and overlaps.

For the quantification of the paths curvatures the osculating circle is used, as shown in Fig. 2. The curvature κ measures how sharply a path bends at a specific point P and is defined as the reciprocal of the radius r of the osculating circle at this point.

$$\kappa = 1/r \tag{1}$$

The osculating circle is defined as the circle at a point P and its tangent that locally approximates the curve at this specific point.

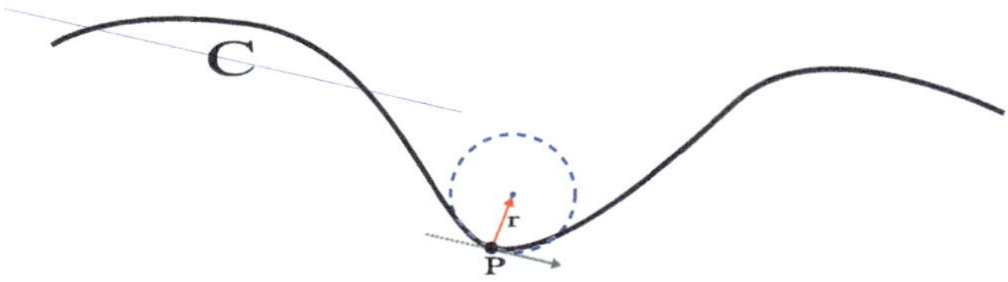

Fig. 2. Schematic of the osculating circle (blue) of a curve C (black) at point P.

3.2 Geometry-Constrained Path Planning Approach

At first the principal stress direction of the assumed load case is extracted from the results of a FEA. Since the resulting vector field can be distributed turbulently and thus lead to manufacturing defects and harsh curvatures [16], a smoothing has to be conducted in order to get appropriate results in the subsequent path planning step.

For this, the extracted turbulent stress direction vector field is smoothed using an inverse distance weighting. Here, each stress direction vector i is smoothed with the stress directions of the neighboring elements j weighted with the inverse of the squared Euclidean distance to the considered element d_{ij}:

$$w_{ij} = \frac{1}{d_{ij}^2} \tag{2}$$

In order to form load-path-adapted fiber paths based on the smoothed vector field, elements that align with the vector field must be associated in order to form a path. The path generation algorithm utilises a geometry-constrained approach, whereby a conical shape is employed to identify suitable successor points that form a path.

The algorithm is designed to generate paths through a set of points, the finite element center points, by identifying connections along the principal stress direction based on directional and spatial proximity. It works iteratively to construct these paths while ensuring geometric constraints, which are in compliance with the manufacturing requirements.

Starting with an initial point, the algorithm considers successors in its associated smoothed stress direction and evaluates neighboring points that fall within a predefined range. This range is represented as the volume of a constructed cone defined by its height h and radius r, with the initial point as the apex and its stress direction vector defining the cone's orientation. Points that lie inside this cone are selected as potential candidates for the next point in the path. In the Fig. 3 the construction of a cone to find potential successor points is depicted.

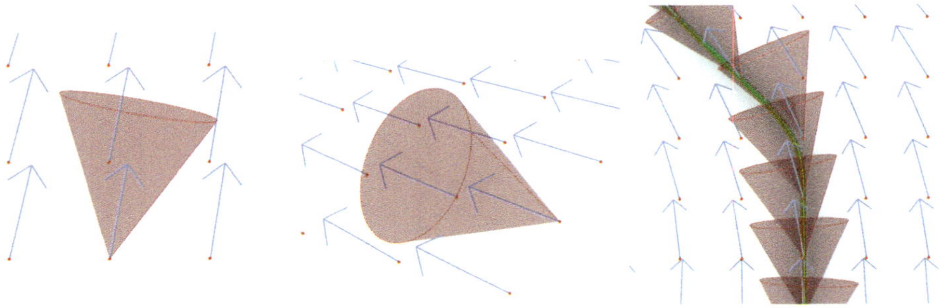

Fig. 3. Constructed cone along the stress direction vector for finding potential successor points for creating a load-path-adapted paths.

The steering radius of the generated paths can be controlled indirectly by adjusting the cone parameters. Once a suitable successor point is found, it is added to the path and becomes the new starting point, and the process is repeated. The vector field smoothing that has been conducted in advance, in conjunction with the geometric dimensions of the cones in the direction of the load, serves to prevent abrupt and sharp changes in direction, which would otherwise result in a kinked path. This iterative approach allows the algorithm to build a continuous path, following the general direction of the principal stresses. If no further points are found within the range, the current path ends, and the algorithm moves on to start a new path with any remaining points.

The resulting paths are organized hierarchically, grouping points that belong to the same path. The algorithm ensures paths meet the minimum deposition length to avoid incomplete or non-manufacturable paths. Additionally, any points that cannot be included in a path are recorded as unused, in order to run further analysis e.g. coverage analysis.

The identification of the points constituting a given path is followed by the establishment of a connection between these points by means of a non-uniform rational B-spline, for which the aforementioned points are used as control points.

The curvature of each curve along the path is then calculated using the osculating circle. The curvature is compared to a given minimum radius that is not allowed to be undercut in order to comply with the material's minimum steering radius needed for a defect-free deposition. If the constraint of the minimum steering radius is violated, the cone parameter must be adjusted, so that the constraint is fulfilled.

4 Summary

This paper introduces a novel concept for path planning in AFP, integrating load-path-adapted fiber placement with geometric and directional constraints. By utilizing smoothed vector fields derived from FEA results, the proposed algorithm aims to generate fiber paths that align with principal load directions while ensuring manufacturability. The method leverages a geometry-constrained approach using adjustable cone parameters to control critical deposition parameters, such as minimum steering radius, to minimize defects.

A key advantage of this approach is the simplicity, as it relies on only two adjustable variables – the cone height and cone radius – to generate fiber paths. This reduces computational complexity compared to optimization-based methods, which typically involve a high number of design variables. Despite its simplicity, the method is designed to ensure defect-free deposition while maintaining structural performance.

As this work presents a conceptual approach, the proposed method has not yet been tested on specific cases or real-world geometries. However, based on theoretical considerations and the integration of known manufacturing constraints, it is expected to provide a viable and efficient solution for generating curvilinear fiber paths. Future research will focus on implementing and evaluating the method on test cases, validating its effectiveness in practical applications, and assessing its impact on structural performance. Furthermore, integrating this approach into optimization frameworks could enhance its applicability in lightweight and high-performance composite structures.

Acknowledgements. This research paper in project LaiLa is funded by dtec.bw Digitalization and Technology Research Center of the Bundeswehr which we gratefully acknowledge. dtec.bw is funded by the European Union NextGenerationEU.

References

1. Dorey, G.: Carbon fibres and their applications. J. Phys. D Appl. Phys. **20**(3), 245 (1987). https://doi.org/10.1088/0022-3727/20/3/002
2. Fleischer, J., Teti, R., Lanza, G., Mativenga, P., Möhring, H.C., Caggiano, A.: Composite materials parts manufacturing. CIRP Ann. **67**(2), 603–626 (2018). https://www.sciencedirect.com/science/article/pii/S0007850618301537
3. Zhang, W., Liu, F., Jiang, T., Yi, M., Chen, W., Ding, X.: Overview of current design and analysis of potential theories for automated fibre placement mechanisms. Chin. J. Aeronaut. **35**(4), 1–13 (2022)

4. Hao, P., et al.: Design of manufacturable fiber path for variable-stiffness panels based on lamination parameters. Compos. Struct. **219**, 158–169 (2019)
5. Brasington, A., Sacco, C., Halbritter, J., Wehbe, R., Harik, R.: Automated fiber placement: a review of history, current technologies, and future paths forward. Compos. Part C Open Access **6**, 100182 (2021)
6. Heinecke, F., Willberg, C.: Manufacturing-induced imperfections in composite parts manufactured via automated fiber placement. J. Compos. Sci. **3**(2), 56 (2019)
7. Sheikhi, M., Rafiei Anamagh, M., Bediz, B., Tunc, L.T.: Conical-shaped variable stiffness composite laminates: design and fiber path planning. Mater. Des. 113142 (2024)
8. Rousseau, G., Wehbe, R., Halbritter, J., Harik, R.: Automated fiber placement path planning: a state-of-the-art review. Comput.-Aided Des. Appl. **16**(2), 172–203 (2018)
9. Zhang, Z., et al.: Effect of AFP induced triangular gaps on manufacturing quality and stress distribution of composite panels. Compos. Sci. Technol. **243**, 110222 (2023)
10. Tunc, L.T., Sheikhi, M.: AFP tool path planning for manufacture of variable stiffness composites. CIRP Ann. (2023)
11. Khandar Shahabad, P., Rafiei Anamagh, M., Serhat, G., Basdogan, I., Bediz, B.: Advanced lamination parameter interpolation and extrapolation methods for designing manufacturable variable stiffness laminates. Compos. Struct. **326**, 117608 (2023)
12. Rabczuk, T., Song, J.H., Zhuang, X., Anitescu, C.: Chapter one - introduction. In: Rabczuk, T., Song, J.H., Zhuang, X., Anitescu, C. (eds.) Extended Finite Element and Meshfree Methods, pp. 1–18. Academic Press (2020). https://www.sciencedirect.com/science/article/pii/B9780128141069000073
13. Tian, Y., Shi, T., Xia, Q.: Optimization with manufacturing constraints for composite laminates reinforced by curvilinear fibers through a parametric level set method. Compos. Struct. **321**, 117310 (2023)
14. Arsenyeva, A., Duddeck, F., Thompson, H.M.: An ISO-contour method for automated fiber placement optimization of composite structures. Compos. Struct. 117628 (2023)
15. Albazzan, M.A., Harik, R., Tatting, B.F., Gürdal, Z.: Efficient design optimization of nonconventional laminated composites using lamination parameters: a state of the art. Compos. Struct. **209**, 362–374 (2019)
16. Kipping, J., Schüppstuhl, T.: Load-oriented nonplanar additive manufacturing method for optimized continuous carbon fiber parts. Materials **16**(3) (2023)

Influence of a Novel Fixture Hardening Procedure on the Tooth Root Load Carrying Capacity of Case-Hardened Gears

Yves Johannes Barth[1](✉), Sven Wagner[2], Martin Hunkel[3], Thomas Tobie[1], and Karsten Stahl[1]

[1] Gear Research Center (FZG), Technical University of Munich, 85748 Garching, Germany
yves.barth@tum.de

[2] HEESS GmbH & Co KG, 68623 Lampertheim, Germany

[3] Leibniz-Institut für Werkstofforientierte Technologien - IWT, 28359 Bremen, Germany

Abstract. During the design of electric vehicles, significant efforts are being made to identify and implement lightweight construction potentials to enhance efficiency. The electric motor is a major weight driver due to its large dimensions, necessary to meet the high torque demands at low speeds. To address this drawback, new drive concepts are being developed that utilize smaller electric motors operating at high speeds with reduced torque. These concepts require the synchronization of the motor's high speeds with the drivetrain through a reduction gearbox. The high centrifugal forces of high-speed drivetrains impose additional stresses on the gears, resulting in the need for increased material strength.

A well-established process in gearbox manufacturing is fixture hardening, where gears are fixed by mandrels during quenching to minimize component distortion. This paper presents a novel fixture hardening procedure where the forces applied by the fixtures are precisely controlled. By applying targeted forces during the martensitic transformation, the load carrying capacity of the material can be significantly improved. Pulsator tests on tooth root load carrying capacity for two different materials demonstrate the potential of this process for case-hardened gears. It is shown that the developed fixture hardening process offers a simple and easily integrable solution for enhancing the load carrying capacity of gearbox components in existing production lines.

Keywords: Gears · Lightweight Design · Fixture Hardening · Tooth Root Load Carrying Capacity

1 Introduction

During the design of electric vehicles, significant efforts are being made to identify and implement lightweight construction potentials to enhance efficiency. The electric motor is a major weight driver due to its large dimensions, necessary to meet the high torque demands at low speeds. To address this drawback, new drive concepts are being developed that utilize smaller electric motors operating at high speeds with reduced torque.

F. Endress et al. (Eds.): MSLD 2024, *Proceedings of the Munich Symposium on Lightweight Design 2024*, pp. 9–21, 2025.
https://doi.org/10.1007/978-3-032-01360-6_2

These concepts require the synchronization of the motor's high speeds with the drivetrain through a reduction gearbox. The high centrifugal forces of high-speed drivetrains impose additional stresses on the gears, resulting in the need for increased material strength. To support the transformation of the industry towards electrical mobility, the methods to increase the material strength should be easy to implement into existing facilities.

1.1 Fixture Hardening

In addition to free quenching of components in batches, where the parts are generally subject to distortion, quenching in hardening presses (fixture hardening) is a well-established practice in the manufacture of low-distortion case-hardened components [17, 19, 21]. According to the current state of research, this production method can be used to produce a component with moderate residual compressive stresses in the surface layer and minimal dimensional and shape changes.

Fixture hardening is characterized by a tool (fixture) which is integrated in the quenching machine, into which the heated parts are inserted individually. The quenching is done under application of transient individually controlled forces on the mandrel and outer down holder and via a controlled cooling media flow. An exemplary setup for a spur wheel is shown in Fig. 1. The gear is austenitizied and carburized outside of the fixture hardening machine in a separate furnace. After the carburizing, the gear is placed on the mandrel and fixated between the down holders and the base plate. Once the gear is fixated, the quenching medium is injected into the machine under pressure. After the quenching is finished, the gear is ejected automatically.

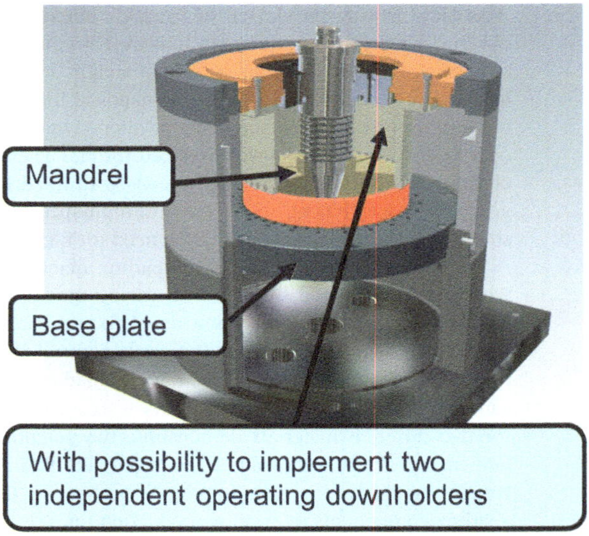

Fig. 1. Standard fixture hardening tool

The control of hardening distortion by means of fixture hardening devices is facilitated in case-hardened components by their usual dimensional change behavior during quenching. As a result of the carbon graded surface layer, the transformation in the core, which generally makes up the largest volume proportion of the component, takes place at the beginning and the carburized areas follow the local carbon content as well as the resulting decrease in martensite start temperature from the inside to the outside.

Due to the primary thermal shrinkage of the components during quenching, the component can be shrunk onto a mandrel, where initial reduction of out of roundness displacements takes place. The transformation of the non-carburized core material will cause an increase in volume, which is partially hampered by the austenitic edge layer. Only when the austenitic carburized layer is transformed in the end of the cooling process the components outer dimensions start to increase significantly. A corresponding fixture, can thus limit the dimensional and shape changes [17].

In the case of contracting components, positive results are already achieved with rigidly set fixture mandrels (path-controlled fixtures) onto which the components shrink. If the fixture diameters are large enough, targeted plastic deformation takes place, which keeps both the dimensions and shape of the heat-treated workpiece within narrow tolerances.

If the possibilities of adjustable fixtures mandrels are used, the tool can follow the enlarging workpiece with a defined force (force-controlled fixture) or apply forces selectively during quenching. The applied forces influence the transformation related volume change in a directed manner, and thus the dimensional change behavior via the so-called transformation plasticity, whereby dimensional increases can be achieved to a greater extent and at lower forces applied [10]. To achieve constant finished dimensions, however, these processes must be actively influenced. This requires the possibility of in-situ measurement of dimensional changes and fixture forces. In addition to simply influencing dimensional and shape changes, the use of expanding mandrels and variable external fixings can also be used to introduce stresses at different points during quenching process, which could influence the resulting residual stresses.

1.2 Tooth Root Fracture

Toot root fracture is one of the main gear failure modes and usually results in total failure of a gearbox. The load condition that is decisive for tooth root fracture is the stress from the bending moment that occurs in the tooth root as a result of the power transmission of the gearing. Due to the kinematics of the gearing, this bending moment occurs dynamically, in the case of external gearing usually in a pulsating manner. If the local fatigue strength of the material is exceeded, a crack is initiated, which increasingly grows into the material under further cyclic loading. As soon as this crack reaches a critical length, a violent rupture fracture occurs, which leads to the entire tooth on the gear breaking off. The critical area in the tooth root is usually the 30°-tangent.

While the stress in the tooth root is largely determined by the torque to be transmitted and the geometry of the tooth (lever arm, notch effect in the tooth root) [9], the factors influencing the tooth root load carrying capacity are more numerous and more complex [8, 13, 14, 16].

In the case of conventionally case-hardened gears, recent studies show that the tooth root load-bearing capacity can be described well by the hardness profile, the residual stresses in the surface layer, and the degree of purity of the material [4, 5, 18]. Other influencing factors include the alloy composition, surface condition (e.g., roughness, surface oxidation depth), and microstructure composition [13, 14, 16].

Furthermore, high rotational speeds cause static underloads in the tooth root area. These have a negative effect on the tooth root load carrying capacity [2, 11, 12]. As these centrifugal forces increase with the speed and the radius to the axis of rotation, it is advisable to reduce the mass in areas further out (i.e., the tooth) in order to increase the load carrying capacity of a high-speed gear. Measures that increase the load carrying capacity in terms of lightweight construction, therefore, have a double effect: A stronger structure already allows the size to be reduced. As the reduced size also results in lower centrifugal forces in the gear, the safety against tooth root breakage is further increased.

2 Materials and Methods

2.1 Novel Fixture Hardening Procedure

If plastic deformation occurs before or during a martensitic transformation, e.g., due to external forces, certain parameters influencing dimensional changes are affected. Deformation of metastable austenite impacts the martensite start temperature (M_S) and transformation kinetics [15]. Even applying tensile stresses without plastic deformation increases the M_S-temperature [3]. The potential effects of this dependency on dimensional changes are demonstrated in a computer simulation investigating the influence of varying M_S-temperature on the length changes of quenched cylinders [20].

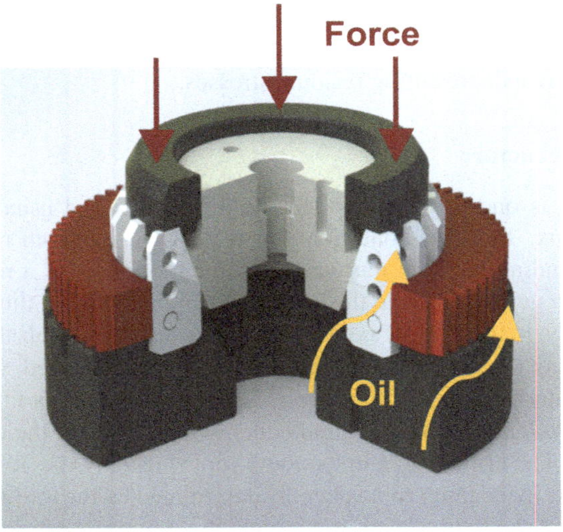

Fig. 2. Novel fixture hardening tool

Martensitic transformation also exhibits transformation plasticity, meaning the dimensional changes align according to the principle of least resistance under mechanical loads [1, 3]. This mechanism enables controlled dimensional changes while maintaining volume constancy.

To enhance residual stresses in the martensitic near-surface layer of a case-hardened component, the component must be elastically stretched immediately after the core transformation to further elastically stretch the still austenitic area near the surface. The subsequent transformation of the area near the surface will lead to a directed martensitic transformation, relaxing the additional stresses by additional straining of the surface.

Once the stretching is relieved by releasing the mandrel from its tension, the additional elastic spring back is likely to result in increased compressive stresses in the surface layer.

In order to be able to apply targeted strain during the martensitic transformation, a suitable tool first had to be developed for the novel fixture hardening procedure (see Fig. 2). Instead of a fixed mandrel, a mandrel consisting of several lamellae is used. These lamellae lie inside the bore of the gear wheel and are activated via a conical punch. Grooves in the tool enable the controlled flow of the quenching medium around the component.

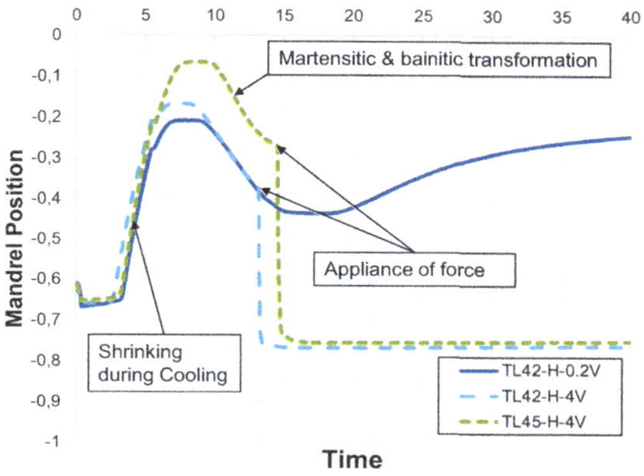

Fig. 3. Mandrel position during quenching

The position of the punch is constantly monitored and thus allows conclusions to be drawn about the transformation behavior of the component. Figure 3 shows a qualitative example of the cone position of three components during quenching. In the first section, the component initially shrinks linearly due to the decrease of temperature and resulting thermal shrinkage of the austenite. At a certain point, the transformation of the core begins, which slows down the shrinkage of the component and ultimately leads to an increase in volume. The curve for TL42-H-0.2V shows the further cooling behavior for a component without external force. The martensitic transformation subsides at some point and renewed shrinkage occurs due to a decrease in temperature. With TL42-H-4V and TL45-H-4V, on the other hand, a controlled force is applied by the mandrel as soon as the transformation of the core is complete. The cone remains in a fixed position and thus introduces targeted strain into the component. The elastic spring back occurs during the ejection of the parts and is not shown in Fig. 3.

It is crucial to time the completion of the core transformation before the surface transformation, as elastic stretching can only be introduced at this stage and subsequently converted into directed transformation strain of the surface.

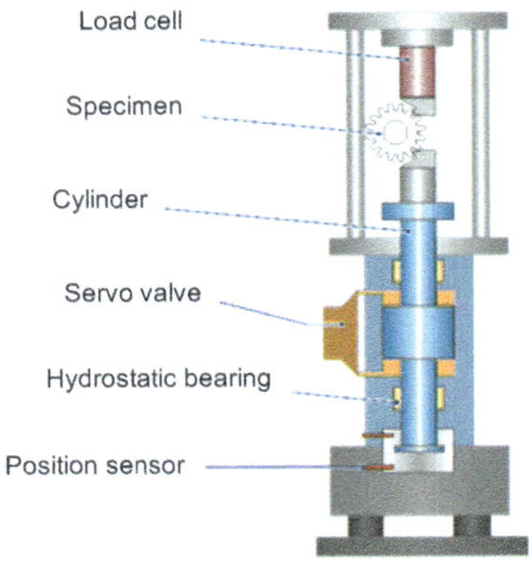

Fig. 4. Hydraulic pulsator test rig

2.2 Test Setup

In this study, a hydraulic pulsator test rig was used to determine partial S-N-curves for tooth root bending strength of the test gears (see Fig. 4). The pulsator test rig mainly consists of a hydraulic cylinder with a servo valve and a jaw attached to it. A second jaw is attached to a load cell. The jaw geometry is designed to resemble the stiffness of a mating gear. The test gear is then fixed between the two jaws, and the contact pattern is adjusted. Via the servo valve the cylinder applies a pulsating force excitation on the test gear. The force and position of the cylinder are constantly measured and controlled. The pulsator test rig used in this study has a maximum force excitation of 30 kN and a load frequency range of 50–70 Hz.

The S-N curve is determined in two sub-tests each. The endurance limit is determined using the improved staircase method according to Hück [7]. First, an initial load level is determined based on empirical values. Then the clamped pair of teeth is subjected to a pulsating load until it breaks, or a predefined number of load cycles (in this study: 6 million load cycles) is reached. If a tooth breaks, the test point is considered a failure, and the load for the next test point is reduced by a defined step (in this study: 0.25 kN). If the test point reaches the maximum load cycle number without damage, the test point is evaluated as a runout, and the load for the next point is increased by one step. Figure 5 shows an example of a test assignment. To obtain a valid staircase, there must be at least one pure failure level with at least two test points and one pure runout level. Failures above the lowest pure failure level and runouts below the highest pure runout level are not considered in the evaluation. The artificial test point at the end of the staircase step contains the information as to whether the last real test point is a failure or a runout.

With a fully occupied S-N curve, the number of test points for the endurance limit is 10–12 test points. For a partially occupied S-N curve, the number is reduced to 6–8 test points. Following the test, the pulsator normal force $F_{Pn\infty,50\%}$ for 50% probability of failure is calculated from the staircase and converted into the tooth root nominal stress

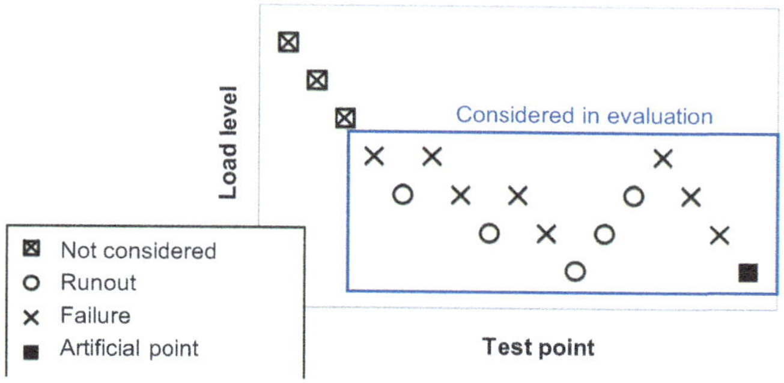

Fig. 5. Staircase method acc. to Hück [7]

$\sigma_{F0\infty,50\%}$ using the formula according to ISO 6336-3 [9]:

$$\sigma_{F0\infty,50\%} = \frac{F_{Pn\infty,50\%} \cdot \cos\alpha_n}{b \cdot m_n} \cdot Y_F \cdot Y_S \cdot Y_\beta \tag{1}$$

$\sigma_{F0\infty,50\%}$ in N/mm^2	Tooth root nominal stress for 50% failure probability
m_n in mm	Normal modulus
$F_{Pn\infty,50\%}$ in N	Pulsator normal force for 50% failure probability
Y_F	Form factor
α_n in deg	Normal pressure angle
Y_S	Stress correction factor
b	in mmWidth of the tooth
Y_β	Helix angle factor (= 1 for spur gears)

To determine the finite fatigue strength, the procedure according to Haibach [6] is used. Two load levels above the endurance limit are each assigned 3–5 test points (F_{Pn1} and F_{Pn2}). For each load level, the geometric mean of the number of cycles endured by the individual test points is determined ($N_{1,50\%}$ and $N_{2,50\%}$). Assuming a logarithmic normal distribution, the slope exponent k of the S-N curve in the finite fatigue range can be determined according to formula (2). The conversion of the pulsator normal forces into the nominal tooth root stress is analogous to formula (1). The complete S-N curve results from the superposition of the test results for the endurance limit with the test results for the finite fatigue strength.

$$k = \frac{\log(N_{2,50\%}) - \log(N_{1,50\%})}{\log(F_{Pn1}) - \log(F_{Pn2})} \tag{2}$$

k	Slope exponent
F_{Pni} in N	Pulsator normal force level
$N_{i,50\%}$	Geometric mean number of failure load cycle for each load level

Table 1. Gear geometry

Parameter	Sign	Value
Modulus	m_n	1.65 mm
Number of teeth	z	81
Normal pressure angle	α_n	15°
Tip diameter	d_a	137.7 mm
Profile shift coefficient	x^*	0.296
Tooth width	b	18 mm

Table 2. Test variants

Variant	Description
TL42-H-0.2V	Reference variant (min. Force)
TL42-H-4V	Highest force application
TL45-H-4V	Highest force application + different material

2.3 Gear Geometry and Test Variants

The gear geometry in this study is representative for the wheel of the first stage of a 2-stage transmission in a modern electric vehicle. The main geometry parameters are given in Table 1.

Three different variants were the subject of this study (see Table 2). TL42-H-0.2V is a case-hardened reference variant made of TL4227 steel (variant of the 20MnCr5 steel grade) without additional force application during quenching. TL42-H-4V is a variant of the same steel batch with maximum force application after completion of the core transformation and TL45-H-4V with maximum force application and variation to the material TL4521 (variant of the 20NiMoCr6-5 steel grade). All gears were carburized at 950 °C for 1 h.

In the pulsator test rig, the number of teeth in the gap between the jaws influences the position of the jaws relative to the tooth profile and, therefore, the form and stress correction factors Y_F and Y_S. In this study, a total number of 8 teeth was in between the jaws. To account for possible differences in the tooth root geometry due to manufacturing tolerances and the novel fixture hardening procedure Y_F and Y_S are determined separately for each batch of test gears.

The values for Y_F and Y_S, as well as the surface roughness representative for the tooth root and the working pressure angle of the pulsator normal force, are given in Table 3. The variance of the characteristic values determined is within the normal range within a production batch. An influence of the novel procedure is not determinable.

Table 3. Variant Documentation

For 8 teeth clamping	Sign	TL42-H-0.2V	TL42-H-4V	TL45-H-4V
Working pressure angle	α_{Fe}	15.6°		
Form factor	Y_F	1.76	1.71	1.65
Stress correction factor	Y_S	1.89	1.91	1.98
Surface Roughness	Rz	3.29 µm	4.23 µm	2.88 µm

3 Results and Discussion

3.1 Metallographic Analysis and Residual Stress Measurement

Hardness and residual stresses were measured for the reference variant TL42-H-0.2V and the variant TL45-H-4V. The hardness measurement of TL42-H-0.2V is also representative for TL42-H-4V since the material batch and the main heat treatment parameters are the same. The hardness in the tooth root is given in Fig. 6. The curves for the TL42-H-0.2V and TL45-H-4V are very comparable. The case hardness depth CHD_{550} is ≈ 0.25 mm for both variants, therefore matching the recommendation of $0.15 \cdot m_n$ acc. to ISO 6336 [8].

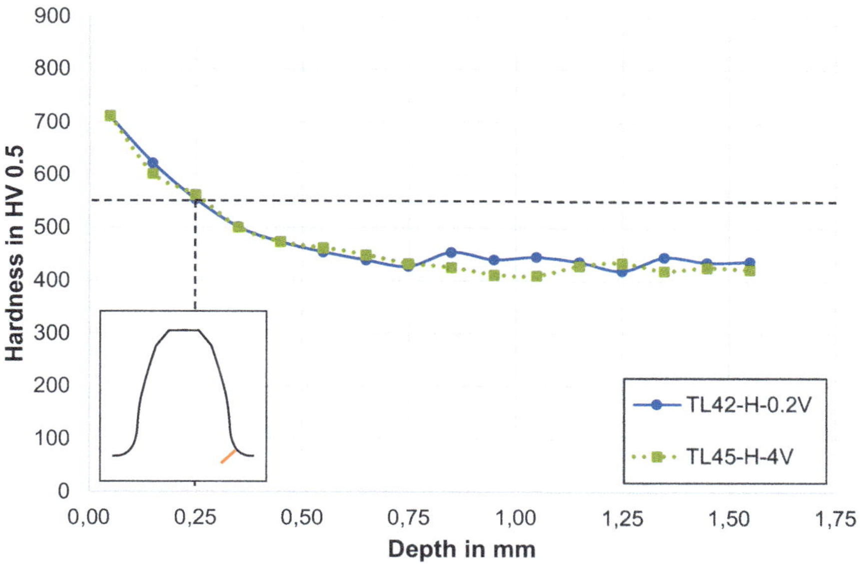

Fig. 6. Hardness at the 30°-tangent

Extensive residual stress measurements were performed in the area of the tooth root. Measurements were conducted in axial and radial direction on 5 teeth of each variant.

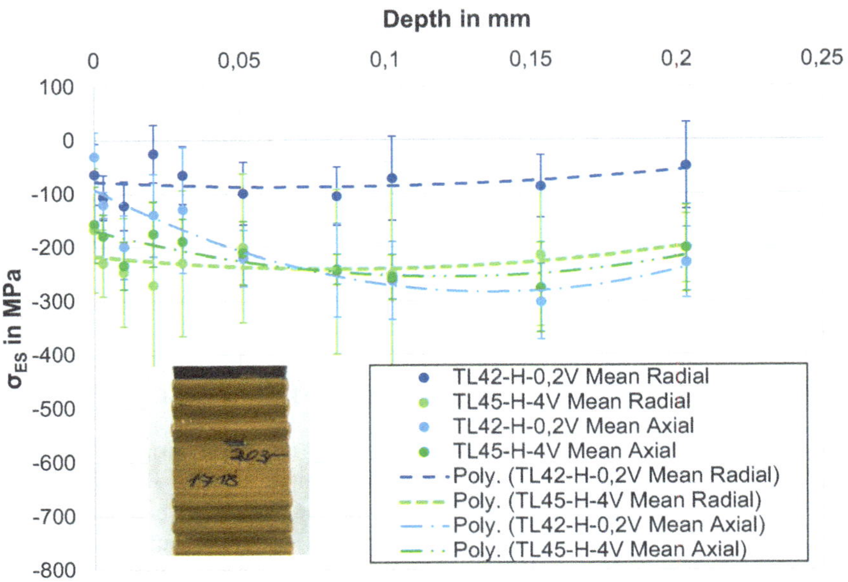

Fig. 7. Residual stresses in the tooth root

Since it is assumed that elastic stresses are present, the gear rim was left intact during the measurements, resulting in a complex measurement setup. To ensure free access to the measuring location in the tooth root, the teeth in front the measuring tooth were removed by erosion (see Fig. 7) instead of cutting small samples out of the gear and the measurement device had to be fitted with a special holder for large specimens.

The residual stress curves given in Fig. 7 are second-degree polynomial fittings through the mean values of the 5 teeth for each variant. The error bars indicate the variance of the values for each tooth as well as the measurement uncertainty for each measurement point. It can be observed, that the measured residual stresses are compressive in the whole measured area. The axial residual stresses for both variants are comparable, whereas the mean radial compressive residual stresses for the TL45-H-4V variant with the highest force application during quenching are ≈ 140 MPa higher compared to the TL42-H-0.2V reference variant with minimal force application.

3.2 S-N-Curves

Figure 8 shows the S-N-Curves for the nominal tooth root stress obtained during pulsator testing and an exemplary image of a fracture surface. The fracture surface shows a symmetrical growth of the fatigue crack with crack initiation in the region of the left and right face side (smoother area). The rougher area characterizes the violent rupture failure. All cracks that led to failures during the pulsator tests initiated at the surface.

The values for the endurance limit $\sigma_{F0\infty,50\%}$, the number of load cycles at the transition point between finite life fatigue and endurance limit $N_{D,50\%}$ and the slope exponent of the finite life fatigue boundary $k_{50\%}$ are given in Table 4.

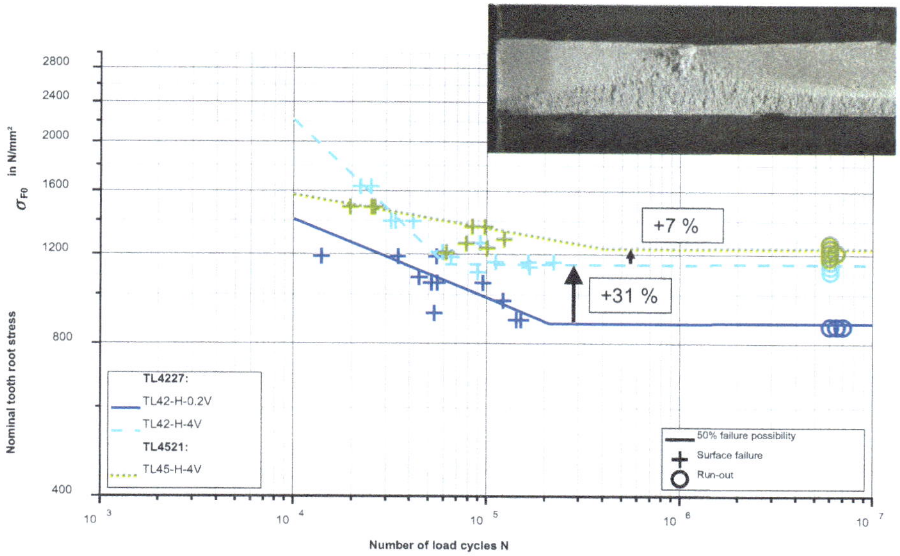

Fig. 8. S-N-Curves for the nominal tooth root stress and crack surface

Table 4. S-N-Curve Data

50% failure probability	Sign	TL42-H-0.2V	TL42-H-4V	TL45-H-4V
Endurance limit	$\sigma_{F0\infty,50\%}$	$880\frac{N}{mm^2}$	$1151\frac{N}{mm^2}$	$1237\frac{N}{mm^2}$
Transition load cycle number	$N_{D,50\%}$	210446	60820	412447
Slope exponent	$k_{50\%}$	6.5	2.7	15.2

3.3 Discussion

The endurance limit $\sigma_{F0\infty,50\%}$ of the TL4227 variant, that was strained during quenching, TL42-H-4V increased by 31% compared to the conventional quenched reference variant of the same material batch TL42-H-0.2V. The finite life fatigue performance was also altered and exhibits a steeper slope for TL42-H-4V than for TL42-H-0.2V, therefore making it less susceptible to high overloads. A comparable positive influence on the endurance limit $\sigma_{F0\infty,50\%}$ could also be demonstrated for the steel grade TL4521. Compared to the TL42-H-4V variant, there is a further increase of 7% of the endurance limit for the TL45-H-4V variant with the same quenching parameters. Therefore, the obtained S-N-Curves given in Fig. 8 resp. Table 4 clearly shows the positive effect of the novel fixture hardening procedure on the tooth root load carrying capacity of the tested specimens.

The metallographic analysis shows no significant impact of the novel procedure on the hardness of the specimens (Fig. 6). The compressive residual stresses do not increase in the axial direction but show a tendency to increase in the radial direction (Fig. 7).

Since the tool only applies a force in the radial direction, the observation is consistent with the model concept. Although it is known that higher compressive residual stresses have a positive effect on the load carrying capacity, the (moderate) increase in radial compressive residual stresses observed here is not sufficient enough to fully explain the significant improvement in the fatigue performance of the gears. This indicates that other microstructural properties have been altered that are not detected by the standard metallographic analysis. The analysis of these properties is the subject of future research. Also, further studies to explain the somewhat unusual crack initiation close to the faces sides of the gears have to be done. Overall, however, it can be stated that the results of the metallographic analysis support the results observed in the pulsator test.

4 Conclusion

In this study, a novel fixture hardening procedure has been developed that can be easily integrated into existing systems in series production. It has been demonstrated that the targeted introduction of strain during quenching can have a:

- Positive effect on the tooth root load carrying capacity
- Influence on the slope of the finite life fatigue curve
- Comparable effect on different (case-hardened) steel grades

Therefore, the presented approach provides a simple way to significantly increase the tooth root load carrying capacity of case-hardened cylindrical gears.

Acknowledgments. The presented results are based on the research project "Light4Speed" and supported through the "Technologietransfer-Programm Leichtbau (TTP LB)" by the Federal Ministry for Economic Affairs and Climate Action (BMWK) based on a decision taken by the German Bundestag and supervised by the Project Management Jülich (PtJ).

Gefördert durch:

Bundesministerium für Wirtschaft und Klimaschutz

aufgrund eines Beschlusses des Deutschen Bundestages

Projektträger Jülich
Forschungszentrum Jülich

References

1. Curà, F., Mura, A., Rosso, C.: Effect of centrifugal load on crack path in thin-rimmed and webbed gears. Frattura ed Integrità Strutturale 447–455 (2015)

2. Curà, F., Mura, A., Rosso, C.: Influence of high speed on crack propagation path in thin rim gears. Fatigue Fract. Eng. Mater. Struct. **40**, 120–129 (2016). https://doi.org/10.1111/ffe.12481
3. Forschungskuratorium Maschinenbau: Analytical strength assessment of components - Made of Steel, Cast Iron and Aluminium Materials in Mechanical Engineering: 7th edition. FKM-Guideline. (in German). VDMA-Verlag, Frankfurt a. M (2020)
4. Haibach, E.: Fatigue strength. Method and data for component calculation (in German), 3th edn, revised. Springer, Heidelberg (2006)
5. Hück, M.: An improved method for the evaluation of staircase tests. Materialwiss. Werkstofftech. **14**, 406–417 (1983). https://doi.org/10.1002/mawe.19830141207
6. International Organization for Standardization: Calculation of load capacity of spur and helical gears - Part 3: Calculation of tooth bending strength (ISO 6336–3) (2019)
7. Leonhardt, C., Otto. M FVA-Heft Nr. 1434 - Influence of Press Fits on the Tooth Root Load Carrying Capacity of External Spur Gears (in German), Frankfurt a. M
8. Lewicki, D.: Effect of speed on gear crack propagation direction. In: International Conference on Motion and Power Transmissions (2001)
9. Lewicki, D.: Gear Crack Propagation Path Studies - Guidelines for Ultra-Safe Design (2001)
10. Lewicki, D., Ballarini, R.: Rim thickness effects on gear crack propagation life. Int. J. Fract. **87**, 59–86 (1997)
11. Li, S.: Centrifugal load and its effects on bending strength and contact strength of a high speed thin-walled spur gear with offset web. Mech. Mach. Theory **43**, 217–239 (2007). https://doi.org/10.1016/j.mechmachtheory.2007.01.006
12. Li, S.: Effects of centrifugal load on tooth contact stresses and bending stresses of thin-rimmed spur gears with inclined webs. Mech. Mach. Theory **59**, 34–47 (2013). https://doi.org/10.1016/j.mechmachtheory.2012.08.011
13. Macherauch, E., Wohlfahrt, H.: Residual stresses and fatigue. Fatigue Behavior of Metallic Materials, pp. 237–283 (1985). (in German)
14. Mileti, M., Strobl, P., Pflaum, H., et al.: Design of a hyper-high-speed powertrain for EV to achieve maximum ranges. In: CTI SYMPOSIUM 2018, vol. 119, pp. 265–273 (2018)
15. Pointner-Gabriel, L., Franzelin, T., Morhard, B., et al.: Electric drive units: a set-up for investigating function, efficiency, and dynamics. Vehicles **6**, 1415–1441 (2024). https://doi.org/10.3390/vehicles6030067

Tensile Behavior and Failure Mechanisms of Carbonyl Iron Filled Endless Glass Fiber-Reinforced Epoxy Composite

Josef Ganslmaier[1](✉), Dominik Santl[2], Erik Buchenau[3], Matthias Bleckmann[4], and Philipp Höfer[1]

[1] Institute of Lightweight Engineering, University of the Bundeswehr Munich, 85577 Neubiberg, Germany
j.ganslmaier@unibw.de
[2] Airbus Defence and Space GmbH, 85077 Manching, Germany
[3] Microwaves and Radar Institute, German Aerospace Center, 82234 Oberpfaffenhofen, Germany
[4] Bundeswehr Research Institute of Materials, Fuels and Lubricants (WIWeB), 85435 Erding, Germany

Abstract. Functionalizing polymers by adding micro- or nanoparticles is a method to adjust specific material properties, such as electromagnetic characteristics or thermal conductivity. However, a high filler content can significantly influence the mechanical performance of the material. To address this issue, short- or long-fiber-reinforced plastics (FRP) are used as base material, combining the benefits of particle functionalization with the inherent strength and durability of fiber reinforcement. This approach maintains the desired properties of the material while minimizing the negative impact on mechanical performance. This study explores occurring failure mechanisms and the resulting tensile behavior of carbonyl iron-filled endless glass fiber-reinforced epoxy composite. The underlying influences are investigated through the manufacturing of a test series with varying particle content, ranging from 0 vol.% to 28 vol.%. Tensile tests are conducted according to DIN EN ISO 527. The study shares insights on the trade-off in mechanical performance opposing to increasing electromagnetic absorption properties, for instance for shielding purposes in composite enclosures.

Keywords: functionalized composites · tensile testing · particles

1 Introduction

Polymer functionalization is an approach in materials science that enables the chemical and physical properties of polymer matrices to be tailored to meet specific application needs. By adding functional fillers to a polymer matrix, properties such as thermal conductivity [1], electrical conductivity [2], UV light endurance [3], magnetic characteristics [4], and mechanical strength [5] can be

F. Endress et al. (Eds.): MSLD 2024, *Proceedings of the Munich Symposium on Lightweight Design 2024*, pp. 22–34, 2025.
https://doi.org/10.1007/978-3-032-01360-6_3

significantly altered. These modifications enable new advanced applications in aerospace, electronics, medicine, and energy systems. Such fillers are often particles in various shapes and sizes in the nanometer and micrometer range.

By adding ferromagnetic particles for example iron [6], certain configurations of iron alloys (ferrosilicon FeSi) [7] or iron oxides (magnetite Fe_3O_4) [8], magnetic properties with electromagnetic dissipation mechanisms can be added to a polymer, allowing the application of the material as an electromagnetic wave absorber. Research has shown that the type, concentration, and dispersion of magnetic fillers play a critical role in determining the effectiveness of absorption. A high volume of fillers is often required to significantly influence the magnetic properties using particles [9]. This can lead, among other parameters such as particle size and particle-matrix interface adhesion, to a notable change in the mechanical performance of the composite system [10]. Poor particle-matrix adhesion can lead to stress concentrations, resulting in reduced tensile strength and toughness. Surface treatments and functionalization of the ferromagnetic particles have been investigated to improve their compatibility with the polymer matrix [11]. The uniform dispersion of particles is difficult to achieve due to agglomeration tendencies, and the processing of highly filled systems can complicate manufacturing processes.

Functional particles can be added to endless fiber-reinforced composites to achieve high mechanical performance and desired electromagnetic properties. An established method involves adding the fillers to the resin system and subsequently infiltrating the semi-finished fiber product in a resin transfer moulding (RTM) process. According to Nordlund et al. [12] this approach is disadvantageous for filler materials with micrometer-sized particles, since strong filtration effects occur on closely packed fibers. This leads to an inhomogeneous dispersion of the filler, which induces a spatial dependency of the permittivity and more complex material behavior. In addition, the formation of particle agglomerates and voids leads to reduced reliability and, in general, decreased mechanical performance.

The chosen strategy within this research is to add the fillers directly to a dry glass fiber roving, to minimize the risk of inhomogeneity. This allows the roving to be processed into any semi-finished product using conventional manufacturing methods for glass fiber-reinforced plastics (GFRP). In particular, this opens up the possibility of producing semi-finished products like prepregs that enable precise control over fiber orientation, resin content, particle content, and thickness, resulting in high-performance composite materials with superior mechanical properties and consistent quality.

In this study, the influences of adding different quantities of carbonyl iron powder (CIP) on the composition and the tensile properties of GFRP are investigated through a testing series using DIN EN ISO 527-5. In addition, separate tests in resin matrix samples according to DIN EN ISO 527-2 are conducted. In total two pure matrix configurations and five GFRP configurations with 0° fiber orientation and varying filler contents are tested. To analyze the trade-off between the mechanical and electromagnetic properties, electromagnetic tests are performed using a waveguide test setup.

This research is focused on spherical CIP in a tensile load scenario. However, other particle sizes and shapes are also relevant for understanding the material behavior. In related work by Santl et al. [13], the incorporation of FeSi in different sizes and shapes is investigated.

2 Experimental Procedure

2.1 Materials

The epoxy system was screened for low viscosity at a processing window of 40–60 °C for thorough infiltration of the rovings. A towpreg epoxy was chosen consisting of 4 parts per weight Accelerator 1573, 28 parts per weight Aradur®1571, and 100 parts per weight Araldite®LY 3508 (Supplier: Huntsman Corporation). The blended system shows a viscosity of 2000–2200 mPas.

S-Glass fibers were chosen due to their high tensile strength. The specimens were produced using HMC9-272 T4(68) (Supplier: Oschatz Glasseiden GmbH). The roving has 272 tex and the glass filaments have a diameter of 9 µm.

The filler material used is spherical carbonyl iron powder (CIP). Ferromagnetic particles such as CIP represent well-known candidates to adjust magnetic part properties [6–9]. The used configuration is CIP-SQ-R (Supplier: BASF SE) since it has good commercial availability, is anticorrosion coated and has a size distribution with a D50 of 5.7 µm, which was determined to be a suitable size for the manufacturing process in preliminary tests. The distribution density measured using a Microtrac Camsizer X2 is presented in Fig. 1a). Figure 1b) shows the particles recorded with scanning electron microscopy (SEM).

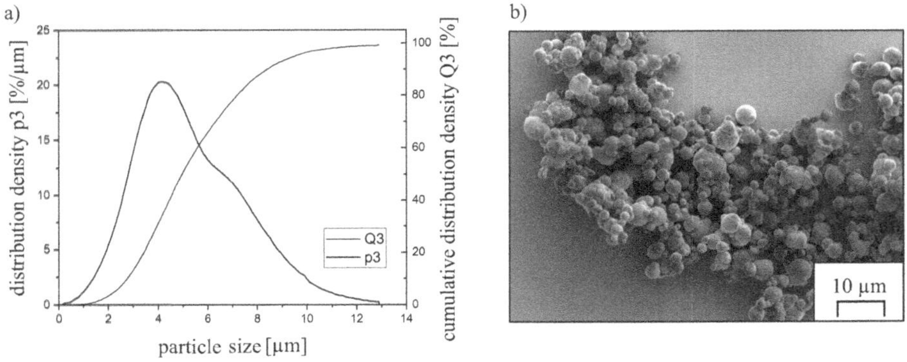

Fig. 1. a) Volume based particle size distribution density p3 and cumulative distribution density Q3 of CIP-SQ-R (D50 = 5.7 µm); b) SEM image of spherical CIP (FOV 80 µm, 1 kV, SE)

2.2 Sample Manufacturing

For the proposed material system, a manufacturing strategy was developed in previous studies. The approach aims to eliminate the spatial dependency of the filler content by integration of the particles at an early stage within the process. As can be seen in the process schematic Fig. 2a), the initial step is to infiltrate the glassfiber roving by a particle-solvent dispersion. In the next step, the remaining solvent on the fibers is dried using infrared lamps, so that a tow with attached particles remains. This tow is then aligned and spread to a fixed width using guiding rollers. The resin is added to the material using a transfer roll. The amount of resin is adjusted to the respective material configuration using a doctor blade and micrometer screws. The impregnation temperature used in this study is 50 °C. In a last step, the tow is wound up on a drum to generate the semi-finished layer (prepreg). The layer thickness is controlled by the winding width on the drum and the targeted material composition. For the samples produced in this study, the layer height is fixed to 0.25 mm.

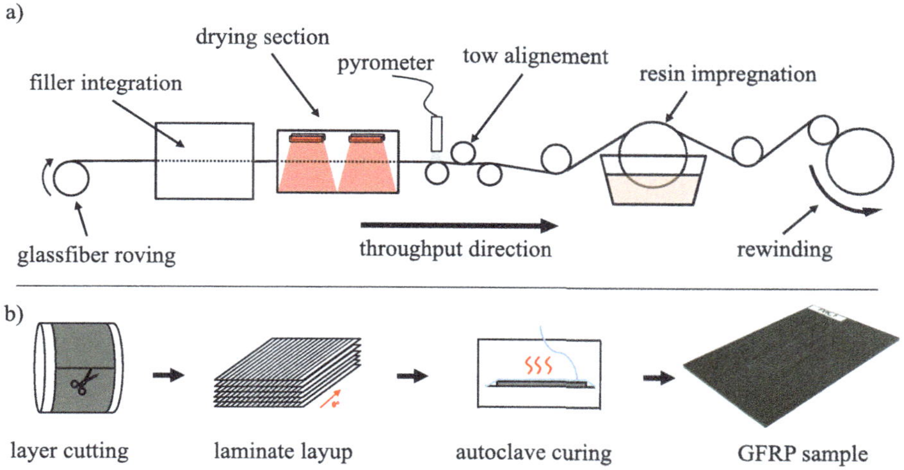

Fig. 2. Process schematic a) manufacturing of semi-finished functionalized GFRP from glass fiber roving to impregnated layer and b) processing sequence from semi-finished layer (prepreg) to GFRP sample

Figure 2b) presents the required steps for producing a sample from the said semi-finished layers (prepregs). This involves a conventional manufacturing approach for fiber-reinforced plastics (FRP). First, the layers are cut to the required size of the sample. The laminate is then stacked and cured within an autoclave for two hours at 120 °C and two hours at 150 °C according to the data-sheet of the used epoxy matrix. To achieve a flat surface without a milling step, a pressure plate is added to the vacuum setup during the curing cycle.

The amount of particles in the material's configuration can be changed through reducing the filler content in the solvent dispersion. To investigate the influence of the composition in the GFRP, five different variants are manufactured by variation of the filler content. In the scope of this study only unidirectional (UD) laminates with 0° fiber orientation are produced.

Pure resin specimens in an unfilled and a filled variant are manufactured and tested to establish direct insights into the effects of the fillers on the matrix. The composition of the filled variant is chosen to match the matrix particle ratio of the highest filled GFRP variant. The samples are mixed and cast in stripes in an aluminum form. The curing is done under vacuum to ensure a mostly non-porous composition. After that the stripes are cut, sanded and then milled to meet the specimen shape Type 5B according to DIN EN ISO 527-2.

2.3 Analysis of the Composition and Structure

With the methodology chosen for the GFRP samples, tracking the exact material composition is difficult since all constituents are added within the enclosed fabrication process. In the curing cycle, excess resin mixed with particles is pressed out of the consolidated sample. Thus, the composition has to be determined at the final sample plate. The chosen methodology employs a standard chemical digestion according to DIN EN 2564 with subsequent mass spectrometry to evaluate the iron content in the probe. The residue is filtered and weighted to determine the mass content of fibers. This method is suitable to establish a local relation between the contents. The homogeneity across the specimen level can only be evaluated if analyses are carried out at several points in the sample, which is very time-consuming.

To investigate the distribution of the fillers within the composite, ion-cross sectional polishes of each configuration are produced. The images were taken using a laser scanning microscope (LSM). To assess the fracture and topological damage, the failure regions after tensile testing are examined using scanning electron microscopy (SEM).

2.4 Tensile Testing

The testing of the GFRP samples is conducted according to DIN EN ISO 527-5 for UD laminates. To ensure correct force introduction, GFRP laminates (±45° fiber orientation) are attached to the test specimen using DELO-DUOPOX AD 895 (Supplier: DELO Industrie Klebstoffe GmbH). A universal testing machine (ZwickRoell GmbH & Co. KG) with a load cell 200 kN and hydraulic clamps 250 kN was used. The strain field is recorded with a video extensometer. For each material variant, five samples with fiber orientation 0° (Type A) are tested.

The pure resin samples are micro tensile tests in accordance to DIN EN ISO 527-2. The used setup was a tensile/compression module with a DDS4 controller by Kammrath & Weiss GmbH. A load cell 200 N is used. The strain rate is

1%/min, resulting in a testing speed of 200 μm/min. At such test speeds, viscoelastic effects in the polymer have to be considered in the tensile behavior of the specimens.

2.5 Electromagnetic Testing

Understanding the interaction of materials with electromagnetic waves is essential for the application as electromagnetic wave absorber. Key parameters such as permeability (μ) and permittivity (ε) provide insights into a material's ability to store and dissipate electromagnetic energy. These properties vary with frequency and are critical for predicting material performance. Various experimental techniques have been developed to measure these parameters accurately.

A waveguide measurement combined with the Nicolson-Ross-Weir (NRW) [14] algorithm is a widely employed method to determine the material constants, particularly the complex-valued permeability (μ) and permittivity (ε), over a range of frequencies. These material constants describe both the storage and loss characteristics of the material for electric and magnetic fields, respectively. They are essential for understanding and predicting the material's interaction with electromagnetic waves, such as reflection, transmission, and absorption behaviors across different frequencies. With the used setup, one frequency band can be studied, using a rectangular material sample precisely machined to fit the waveguide. For this study, a sample of each configuration is sanded to the size required for the X-band (8–12.5 GHz).

The sample is subjected to electromagnetic waves generated by a vector network analyzer. The reflection and transmission coefficients, referred to as S-parameters, are measured and serve as inputs to the NRW algorithm. From these, the relative permeability ($\mu_r = \mu/\mu_0$) and permittivity ($\varepsilon_r = \varepsilon/\varepsilon_0$) are calculated, where μ_0 and ε_0 are the permeability and permittivity of free space, respectively. The complex permeability is expressed as $\mu_r = \mu' - j\mu''$, where μ' the material's ability to store magnetic energy, and μ'' accounts for magnetic losses due to energy dissipation to heat. In this study a ferromagnetic particle material is used, therefore, the focus is on the permeability. This determines the effectiveness of the material or selected particles in attenuating the electromagnetic wave for the desired frequencies.

3 Results

3.1 Material Composition

An overview on the material composition of the GFRP samples is illustrated in Fig. 3a). A variation of the CIP content was successfully produced ranging from 0–28 vol.%. The matrix content increases as soon as particles are incorporated to the system, then remains relatively constant for configurations three and four, and then increases slightly to 50 vol% for configuration five. As can be seen in the cross-sectional polishes in Fig. 3b), the packing density of the fibers is strongly influenced by the presence of particles. This means that complete fiber consolidation can no longer take place, as the particles act as spacers. The gaps are therefore filled with resin. This effect is increasing for higher particle contents at the expense of the fiber content. The relative content of particles purely in the matrix of configuration five is 36 vol.%. This ratio is used for the pure matrix samples.

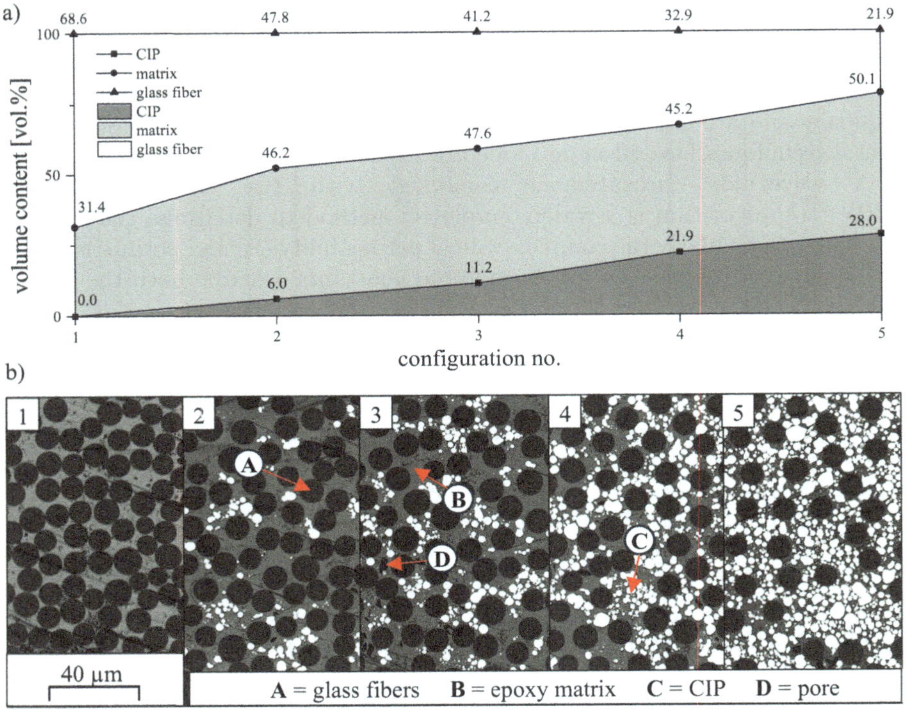

Fig. 3. Material composition of GFRP samples with a) distribution of material proportions CIP, epoxy matrix and fibers; b) LSM micrographs of tested composite variants with varying CIP content

All configurations show an even distribution of glass fibers and no formation of large particle agglomerates. Configuration five with the highest CIP content of 28 vol.% shows some local pockets of higher resin-particle concentration, however they are minor on a macroscopic scale and are not expected to influence the standard deviation in the mechanical data. The micrographs in Fig. 3b) show that smaller porosities remain in the composite. Since these are observed also for the reference specimen, it is likely that they result from the resin impregnation process during manufacturing of the prepreg or the manual laminate layup. No correlation is found between the porosities that occur and the particle volume content. The total amount of pores was found to be low and their influence is assumed to be negligible for the objective of this research.

3.2 Tensile Testing

Tensile strength and Young's modulus are presented in Fig. 4, as they provide information on the material's ability to withstand the applied tensile forces and its stiffness under load. The mean tensile strength decreases from 1556 MPa to 616 MPa, which is a relative decrease of 60%. The Young's Modulus was calculated in accordance to DIN EN ISO 527-1 and decreases by roughly 49% from 56.3 GPa to 28.9 GPa. The standard deviation of both characteristics shows a narrow spread, indicating good homogeneity of the specimens for all configurations.

As stated in the previous Sect. 3.1, with increasing CIP content, the fiber content decreases, which is expected to be the main contributor to the changing mechanical properties. The comparison of the fiber content of reference configuration one and configuration five decreases from roughly 69 vol.% to 22 vol.%, which is a relative decrease of 68%.

No significant influence of the particles on the tensile strength in the GFRP is noted, since the reduction is in a similar range to the reduced fiber content. The Young's modulus shows less reduction, therefore stiffening through the CIP inclusions is conceivable.

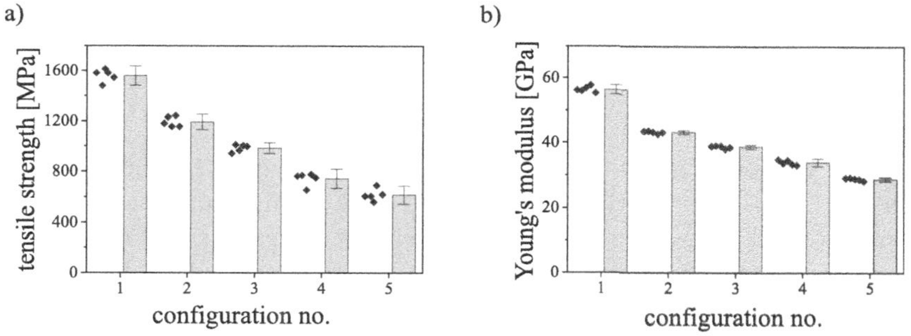

Fig. 4. Mechanical material characteristics from tensile tesing of GFRP configurations according to DIN EN ISO 527-5 with a) the tensile strength b) the Young's modulus

Figure 5a) shows exemplary tensile strain curves of one of the five test specimens of the respective CIP contents. Here it can be seen, that the elongation at break is similar for all configurations, as it is primarily dominated by the fibers. While the reference configuration shows mostly linear elastic properties, increasing nonlinearity in the curves is noted for higher CIP contents. This behavior is pronounced in lower strain regions, thus the previously noted increase in the stiffness is only valid for low strain values. To illustrate this effect, Fig. 5b) shows the Young's modulus evaluated over strains 0–0.5% and 2–2.5%. While the modulus of the reference only drops by 1.8 GPa, configuration five drops by 9.8 MPa, which is a reduction of 3.1% and 35.1% respectively. Effects such as debonding, cracking, and stress redistribution could contribute to the observed behavior. Figure 5c) shows one load and the subsequent relaxation cycle of an additional test specimen from configuration five. The relaxation deviates from the loading cycle indicating permanent damage in the structure of the composite.

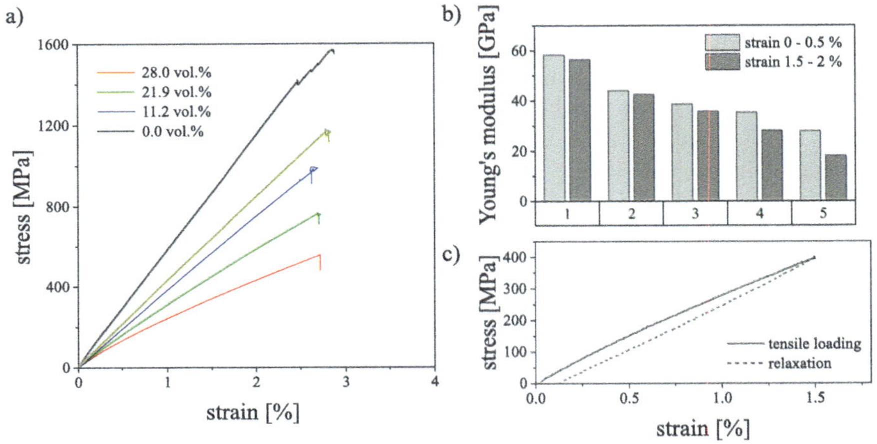

Fig. 5. a) Stress strain curves of exemplary test specimens with denoted volume contents of incorporated CIP; b) evaluation of Young's modulus using strain 0–0.5% and 1.5–2%; c) stress strain load curve to 1.5% strain with subsequent relaxation

In general, reduction in the Young's modulus hints a reduced ability to transfer the tensile load between the different materials. A high mismatch in stiffness promotes local stress concentrations, which in turn can exceed the surface adhesion of the particles to the matrix. After debonding occurred, load transfer efficiency is reduced in the microstructure. This reduces the overall stiffness within the material. This behavior is strongly influenced by the concentration of particles within the matrix. Dekkers et al. [15] demonstrated that at low concentrations of glass beads, the matrix predominantly governs the mechanical response, allowing for relatively efficient stress transfer. However, as the filler concentration increases, local stress concentrations become more pronounced due to particle agglomeration and reduced interfacial bonding strength. These stress concentrations often exceed the adhesion strength at the particle-matrix interface, resulting in debonding and subsequent weakening of the load transfer within the microstructure. This phenomenon ultimately leads to a significant reduction in the composite's stiffness, particularly at higher filler loadings. Additionally, mechanisms such as crazing around the particles, as observed by Dekkers et al. [15], further degrade the material's ability to bear tensile loads.

The results of the pure matrix specimens are presented in Fig. 6. Due to the low strain rate, the unfilled specimens show viscoelastic behavior. This can be seen in the large elongation at constant force in the stress-strain curves Fig. 6a). Therefore, no statement can be made about the quantitative reduction of the elongation at break, but it can be seen that the material looses its ability for viscoelastic deformation as the rigid particles inhibit the matrix mobility. Overall, a reduction to an average elongation at break of 1,3% is noted. No significant influence of the particles is found for the tensile strength in this test series. The Young's modulus increases on average from 3020 MPa for the reference to 9368 MPa for the samples filled with CIP, which is an increase of roughly 210%. The reference modulus is slightly above the material's data-sheet (2900 MPa).

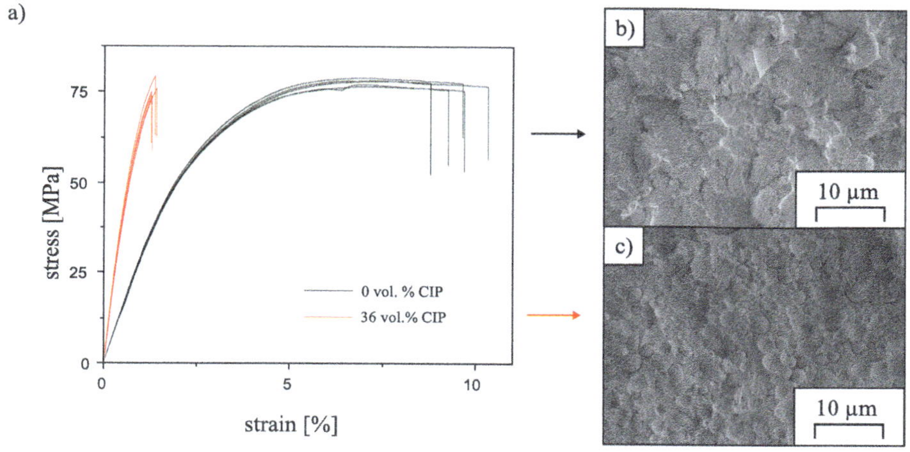

Fig. 6. a) Stress strain curves of particle matrix specimens and SEM images of fracture surfaces of b) exemplary reference sample and c) filled sample (36 vol.% CIP); FOV 50 μm, 1 kV, SE

SEM images of the fracture surfaces of both variants are displayed in Fig. 6b) and c). The particles exposed by the fracture in the variant with embedded CIP show no matrix residue on the surface, from which it is concluded that the crack at failure was along the interface of the two components. The remaining particles consistently show gaps to the matrix. The lost bond indicates that the force transmission is at least partially lost at low strain ranges. As a result, the particles can act as initiators for microcracks or crazing, which explains the significant decrease in elongation at break. Similar debonding topology for CIP embedded in an epoxy matrix is reported for thermally induced stresses in the work of Charles et al. [11].

Transferred to the results of the GFRP samples, it is followed that although no premature failure occurs, a decrease in stiffness is to be expected, as the strain in the GFRP samples well exceeds the elongation at break of the CIP filled matrix. This would manifest itself in mentioned damage to the microstructure, for example in the form of microcracks or crazing. Figure 7 shows a fracture surface in the tensile direction of the GFRP reference and the variant with 28 vol.% CIP. A direct comparison shows that the fracture surface of the filled sample has a more fissured surface. Matrix residues can be seen on the exposed fibers, which generally indicates a good adhesion of the matrix system to the fibers. A closer look at the particles shows that the bond is no longer intact and that the particles cause local splitting in the tensile direction.

Fig. 7. SEM images of fracture surface of GFRP tensile tests extracted from a) destroyed test tab; b), c) CIP content 0 vol.%; d), e) CIP content 28 vol.%, b), d) FOV 200 µm, c), e) FOV 50 µm, 1 kV, SE

3.3 Electromagnetic Characterization

The measured magnetic material properties μ' and μ'' over the X-band are shown in Fig. 8. Both increase depending on the CIP content. The real part shows an increasing dependency of the frequency for higher filler contents. The ability to store magnetic energy diminishes for increasing frequencies, since the magnetic domains cannot realign quickly enough to the oscillating field. The imaginary part remains relatively constant over the frequency. The observed behavior is consistent with the literature [6,9].

The experimental data can be used, for example, as input for the design or simulation of electromagnetic components consisting of the material presented. Overall, it can be seen that the incorporation of different quantities of CIP can notably influence the magnetic properties of an otherwise non magnetic GFRP.

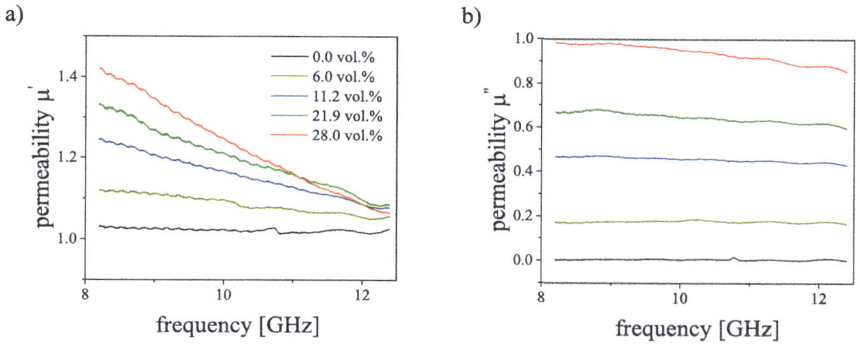

Fig. 8. Complex valued permittivity calculated from wave guide measurement in the X-band (8–12 GHz) of GFRP samples with different volume contents with a) real part μ' and b) imaginary part μ''

4 Discussion and Outlook

In this study, tensile tests were carried out by varying the content of spherical CIP incorporated in GFRP and pure epoxy samples. The analysis of the microstructure showed that the chosen manufacturing strategy is associated with a reduction in the fiber volume content, which dominates the trade-off between the tensile strength and stiffness of the material. The low standard deviations of the GFRP specimens confirm the good homogeneity of the specimens achieved by the manufacturing process. The reduction in Young's modulus due to the reduced fiber content is partially offset by the incorporation of the particles. However, it is found that this strengthening effect is only valid for low strain regions (0–0.5%). In CIP epoxy specimens, the elongation at break is found to be significantly reduced. In the SEM images of the fracture region, the particles show lost boding. It is therefore likely that the matrix in the filled GFRP shows damage in the form of microcracking or crazing prior to ultimate material failure. While this effect does not pose a problem for tensile loading in the fiber direction, it may negatively affect the resistance to aging or impose premature fatigue on the proposed material system. Sufficient bonding of the particles to the matrix is critical. In future investigations, less favorable loading scenarios should be investigated. A promising focus for achieving higher material durability is the functionalization of the particle surface by epoxysilane to achieve better surface adhesion [11].

References

1. Tsekmes, I.A., Kochetov, R., Morshuis, P.H.F., Smit, J.J.: Thermal conductivity of polymeric composites: a review. In: 2013 IEEE International Conference on Solid Dielectrics (ICSD), pp. 678–681 (2013). https://doi.org/10.1109/ICSD.2013.6619698

2. Ozbas, B., O'Neill, C.D., Register, R.A., Aksay, I.A., Prud'homme, R.K., Adamson, D.H.: Multifunctional elastomer nanocomposites with functionalized graphene single sheets. J. Polym. Sci. B Polym. Phys. **50**, 910–916 (2012). https://doi.org/10.1002/polb.23080
3. Rudko, G., Kovalchuk, A., Fediv, V., Chen, W.M., Buyanova, I.A.: Enhancement of polymer endurance to UV light by incorporation of semiconductor nanoparticles. Nanoscale Res. Lett. **10**, 81 (2015). https://doi.org/10.1186/s11671-015-0787-5
4. Wang, Y., Ahmadi Moghaddam, H., Palacios Moreno, J., Mertiny, P.: Magnetic filler polymer composites-morphology characterization and experimental and stochastic finite element analyses of mechanical properties. Polymers **15**, 2897 (2023). https://doi.org/10.3390/polym15132897
5. Coleman, J.N., Khan, U., Gun'ko, Y.K.: Mechanical reinforcement of polymers using carbon nanotubes. Adv. Mater. **18**, 689–706 (2006). https://doi.org/10.1002/adma.200501851
6. Sista, K.S., Dwarapudi, S., Kumar, D., Sinha, G.R., Moon, A.P.: Carbonyl iron powders as absorption material for microwave interference shielding: a review. J. Alloy. Compd. **853**, 157251 (2021). https://doi.org/10.1016/j.jallcom.2020.157251
7. Dosoudil, R., Franek, J., Slama, J., Usakova, M., Gruskova, A.: Electromagnetic wave absorption performances of metal alloy/spinel ferrite/polymer composites. IEEE Trans. Magn. **48**, 1524–1527 (2012). https://doi.org/10.1109/TMAG.2011.2172779
8. Ghigna, T., Zannoni, M., Jones, M.E., Simonetto, A.: Permittivity and permeability of epoxy-magnetite powder composites at microwave frequencies. J. Appl. Phys. **127**, 045102 (2020). https://doi.org/10.1063/1.5128519
9. Chen, P., Liu, M., Wang, L., Poo, Y., Wu, R.: Frequency dispersive complex permittivity and permeability of ferromagnetic metallic granular composite at microwave frequencies. J. Magn. Magn. Mater. **323**, 3081–3086 (2011). https://doi.org/10.1016/j.jmmm.2011.06.061
10. Fu, S.-Y., Feng, X.-Q., Lauke, B., Mai, Y.-W.: Effects of particle size, particle/matrix interface adhesion and particle loading on mechanical properties of particulate-polymer composites. Compos. B Eng. **39**, 933–961 (2008). https://doi.org/10.1016/j.compositesb.2008.01.002
11. Charles, A.D.M., Rider, A.N., Brown, S.A., Wang, C.H.: Improving the actuation performance of magneto-polymer composites by silane functionalisation of carbonyl-iron particles. Compos. B Eng. **196**, 108091 (2020). https://doi.org/10.1016/j.compositesb.2020.108091
12. Nordlund, M., Fernberg, S.P., Lundström, T.S.: Particle deposition mechanisms during processing of advanced composite materials. Compos. A Appl. Sci. Manuf. **38**, 2182–2193 (2007). https://doi.org/10.1016/j.compositesa.2007.06.009
13. Santl, D., Ganslmaier, J., Buchenau, E., Bleckmann, M., Höfer, P.: Functionalization of endless glass fiber-reinforced polymers through incorporating FeSi particles. In: Proceedings of the Munich Symposium on Lightweight Design 2024. Springer (2025)
14. Nicolson, A.M., Ross, G.F.: Measurement of the intrinsic properties of materials by time-domain techniques. IEEE Trans. Instrum. Meas. **19**, 377–382 (1970). https://doi.org/10.1109/TIM.1970.4313932
15. Dekkers, M.E.J., Heikens, D.: The effect of interfacial adhesion on the tensile behavior of polystyrene-glass-bead composites. J. Appl. Polym. Sci. **28**, 3809–3815 (1983). https://doi.org/10.1002/app.1983.070281220

Topology Optimization of Injection Molded Short-Fiber Reinforced Thermoplastic (FRP)-Metal-Hybrids

Sven Lenhardt(✉) and Albert Albers

IPEK – Institute of Product Engineering at Karlsruhe Institute of Technology (KIT), Kaiserstrasse 10, 76131 Karlsruhe, Germany
sven.lenhardt@kit.edu

Abstract. Lightweight design can be used to save resources and reduce greenhouse gas emissions. One way to implement lightweight design is to use short-fiber reinforced thermoplastic (FRP)-metal-hybrids, which combine the advantages of metal and FRP. These can be produced with injection molding by inserting a metal insert into the cavity of an injection molding machine and then over molding it with FRP. However, there is currently a lack of optimization methods to support product developers in the design synthesis of such hybrids.

For this reason, a method for the topology optimization of FRP-metal-hybrids is developed, which aims to increase the pull-out force of the metal insert from the surrounding FRP without damaging the adhesive bond. A weight reduction of the metal insert of such a hybrid structure is prescribed. To ensure that no cavities are created between the FRP and the metal within a sensitivity-based topology optimization, e.g. in Tosca, the method includes an iterative interruption of the topology optimization. The FRP is adapted to the changed structure of the metal insert and the contact pairs between metal and FRP are adjusted. Depending on the structure of the metal insert, the flow paths of the FRP also differ during the injection molding process. For this reason, a mold filling simulation is performed in each iteration of the topology optimization to account for the resulting locally varying anisotropic material properties and residual stresses of the FRP in the simulation model on which the topology optimization is based.

For a metal insert made of S700 and A3WG6 as FRP consisting of polyamide 6.6 with 30% glass fiber reinforcement, different designs result from the topology optimization in Tosca with and without application of the described method. By using the method, designs can be created that avoid premature failure of the cohesive zone. Compared to standard topology optimization in Tosca, higher reaction forces can be achieved without cohesive zone failure.

Keywords: Topology optimization · Hybrid structures · Injection molding

1 Introduction

In view of global warming, it is necessary to reduce greenhouse gas emissions. Lightweight construction takes the approach that less energy is required to move lighter means of transportation and therefore less greenhouse gas emissions are emitted.

F. Endress et al. (Eds.): MSLD 2024, *Proceedings of the Munich Symposium on Lightweight Design 2024*, pp. 35–46, 2025.
https://doi.org/10.1007/978-3-032-01360-6_4

Lightweight construction also offers the advantage of saving resources. To save material of a structural part, high material utilization ratio is necessary.

One way of ensuring high material utilization ratio is hybrid construction. A hybrid structure consists of at least two different materials that contribute to the load-bearing structure. Advantageous properties of the used materials can be utilized in a targeted manner and functional integration is possible [1]. One possible pairing is metal and FRP. The metal on the one hand can be used specifically at force application positions or, for example, at positions in the hybrid structure where multiaxial load cases are present and isotropic material properties are therefore useful. On the other hand, the FRP can be used at positions in the hybrid structure where lower loads are present, or where the force flow is only present in one direction.

Such hybrid structures can be produced intrinsically. This means that when a component is formed, the hybrid is also formed at the same time and therefore no additional time and costs are required for the joining technology [2]. One option for this is injection molding. A metal insert can be placed in the cavity of an injection molding machine, which is then over molded with FRP. In addition to the elimination of the joining production step, such hybrid composites also offer the advantage during recycling that only the melting temperature of the FRP needs to be exceeded to separate the joining partners.

Many factors play an important role in the design of such hybrid structures. One important factor is the adhesive bond between metal and FRP. To improve adhesion, the metal inserts are usually blasted before hybridization, e.g. with corundum [1]. In addition, the fiber orientations in the FRP play a major role, which occur during the mold filling process during injection molding and influence the mechanical material properties. However, mold filling also depends on the geometry of the insert. This leads to the conclusion that the properties of the FRP in the hybrid structure are also influenced by the geometry of the metal insert, which must also be considered in the design process.

The intrinsic production by injection molding provides a high degree of design freedom for the metal insert and therefore complex structures are possible. CAE tools such as structural optimization methods are therefore increasingly being used for the design synthesis of such complex structures. Among these methods, topology optimization offers a great potential for saving weight. With topology optimization, an optimized material distribution can be found in a design space for optimal force transmission even under multiaxial load [3]. However, there is currently no topology optimization method available for such hybrid structures that can consider a cohesive contact interaction and injection molding-specific material properties such as fiber orientation and residual stresses.

2 Optimization Method of FRP-Metal-Hybrids

For the design synthesis of FRP-metal-hybrids, it is necessary to be able to consider the correlation between the structural design and the associated process-specific material properties in the optimization. Since the process-specific material properties cannot be chosen at will but are dependent on the geometry of the metal insert, they should only be considered and not explicitly optimized. For this reason, a coupled optimization method is developed that aims to optimize the interface between metal and FRP by means of a topology optimization of the metal insert, but also takes the process-specific material properties iteratively into account during the design synthesis. Since commercial software tools are widely available and robust, the optimization method is developed based on Abaqus 2022 (structural simulation of the topology optimization), Tosca Structure 2022 (topology optimization framework) and Moldflow 2023 (mold filling simulation). Figure 1 shows the coupling diagram which illustrates the workflow of the method and the software used.

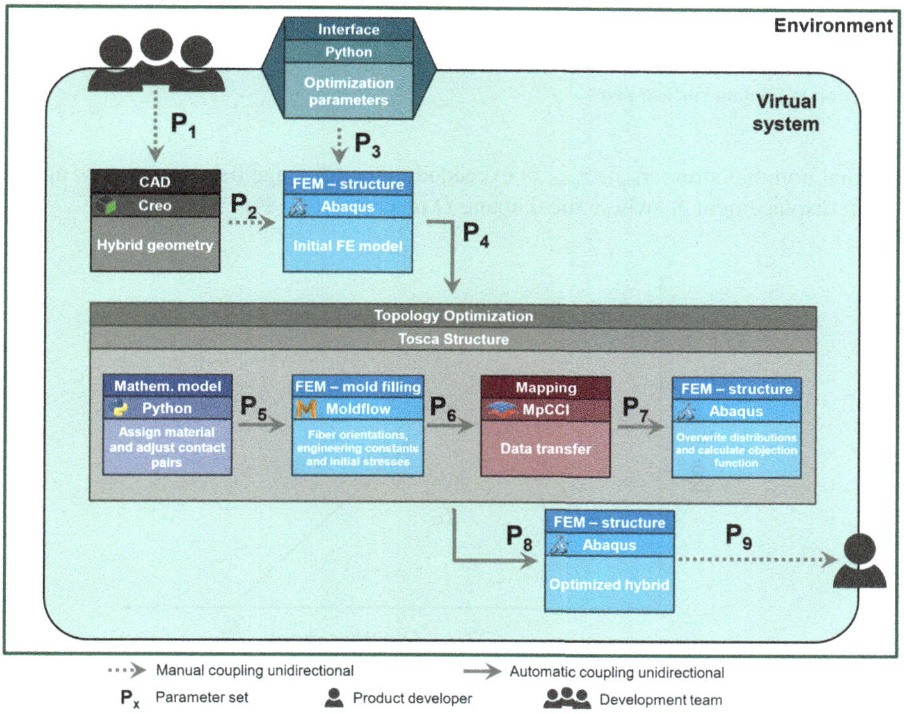

Fig. 1. Coupling diagram with workflow of the developed optimization method for FRP-metal-hybrids, adapted from [4].

First, the product developer, who is part of a development team, creates the CAD geometry of the hybrid structure in the *Hybrid geometry* step, e.g. using Creo Parametric 8. In the next step *Initial FE model*, the geometry is transferred to the finite element (FE) software that serves as the basis for topology optimization, in this case Abaqus. The FE model is created based on that geometry. The interface between the materials is modeled using the cohesive zone contact formulation which has the advantage that no additional elements need to be modelled [5, 6]. The general contact interaction is used to search for contact surface pairs (see Fig. 2).

Normal behavior with hard contact is defined as contact property. Cohesive behavior is defined with stiffness coefficients K_{nn}, K_{ss}, K_{tt}, which are available for the material pairing. Damage initiation and damage evolution are defined as well. As soon as a

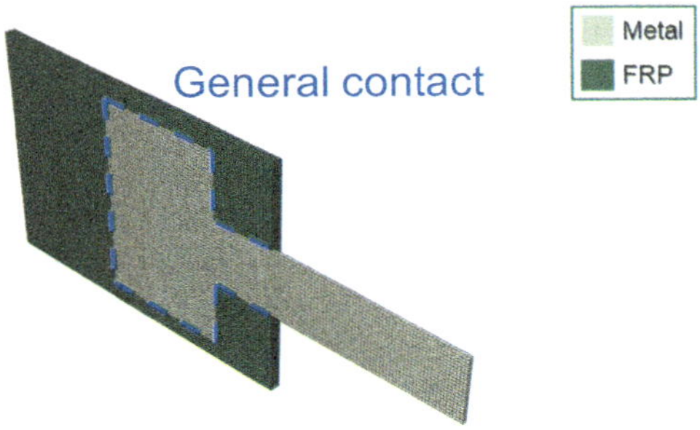

Fig. 2. Cross-sectional view through the midsection of the hybrid structure with general contact interaction at contact surface pairs.

maximal nominal stress σ_{max}, τ_{max} is exceeded, linear damage behavior occurs up to a certain displacement δ_f, where the damage D reaches 1 (see Fig. 3).

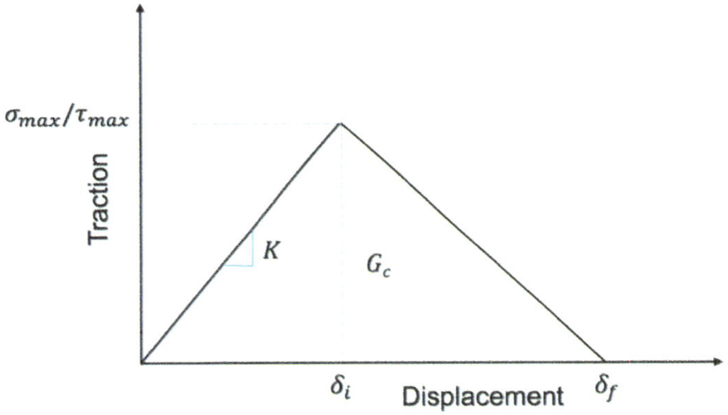

Fig. 3. Traction-Displacement response, based on [7].

The load case always consists of two steps. Before the actual load case, e.g. a specified displacement by pulling out the metal insert and clamping on the FRP side, a step is added. In this step the initial stress distribution of the FRP after injection molding is imported. A warp analysis is performed which results in residual stresses being present in the whole hybrid structure (see Fig. 4). Parameters for the contact, the mesh, the boundary conditions and the load case are set.

The optimization parameters specified by the product developer are integrated in the *Optimization parameters* step via a Python interface. Since the optimizer Tosca Structure is used, which is integrated in Abaqus, and the objective function is to maximize a reaction force (pull-out force of the metal insert), the density-based topology optimization with the sensitivity-based algorithm is selected. The design space lies within the metal insert, as the weight of the heavier component is to be saved. In addition, manufacturing boundary conditions, such as a demolding direction to avoid cavities in the metal, are defined. The

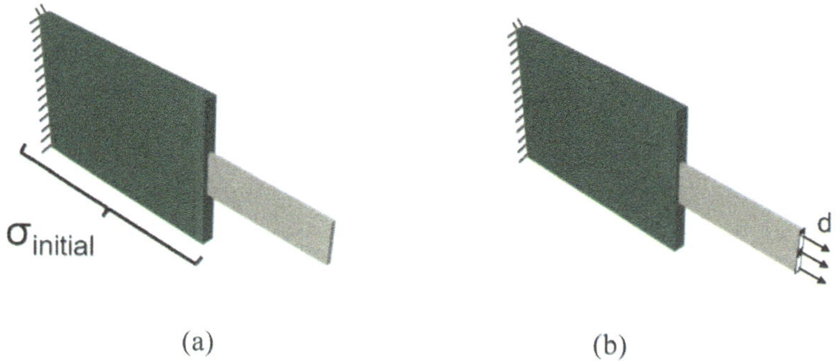

Fig. 4. Load case. Step 1: Residual stress analysis due to initial stresses $\sigma_{initial}$ in the FRP after injection molding (a). Step 2: Pull-out test with prescribed displacement d on the end face of metal insert (b).

optimization is then started. The sensitivities are calculated using Abaqus and adjusted using the method of moving asymptotes (MMA) [8].

Within the topology optimization method, adjustments are made in each iteration (also in the 0th iteration) after the density distribution of the current iteration is available, which are also considered in the FE analysis (FEA) of the iteration. The aim is to achieve a design that is optimally adapted to the process-specific material properties of such hybrids. In the first step of the interruption *Assign material and adjust contact pairs*, the interface between FRP and metal is first determined, e.g. with the iso density value 0.6, which means that all elements with a relative density of less than 0.6 are considered as FRP material. For the contact properties to act on the adapted interface, the elements of the FRP and metal must be decoupled from each other at the interface. This is done by doubling the nodes on the interface and replacing the FRP element nodes with the newly created nodes. This interface is consequently recognized as new contact interface by the general contact interaction. The original interface, which encloses the design space, is adapted with tie constraints to eliminate contacts within the FRP. This is illustrated in Fig. 5.

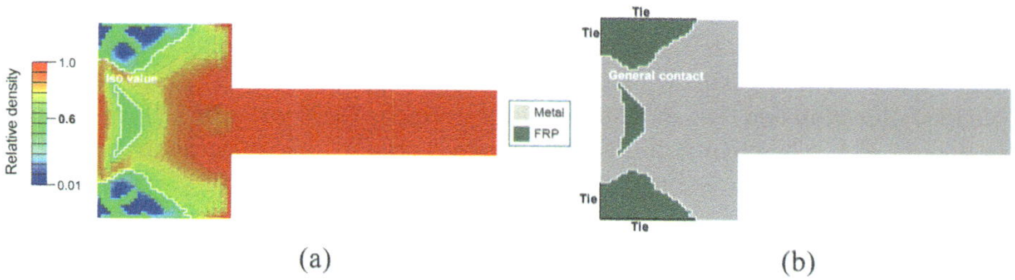

Fig. 5. Interim result with density distribution (a) and material assignment with new contacts (b).

In the next step *Fiber orientations, engineering constants and initial stresses*, the process-specific material properties must be determined for the FRP. For this purpose, a smoothed iso surface of the metal insert is derived in STL format using Tosca Smooth based on the density distribution and the iso value of 0.6. A negative geometry is generated with the geometry of the smoothed metal insert and the cavity of the injection molding machine. This negative geometry is imported to Moldflow. After the FRP material from the Moldflow database is loaded, a mold filling simulation is performed in which the Navier-Stokes equations for non-Newtonian viscosity are solved. To calculate the fiber orientation, Moldflow uses the Folgar-Tucker orientation equation [9]. The 2nd order fiber orientation tensor is created by calculating the probability of fiber orientation. The local material properties are then calculated from the fiber orientation tensor using the Mori-Tanaka model in the form of engineering constants [10]. In addition, the initial stress distribution, which is available directly after injection molding, is exported. Figure 6 shows the results for the fiber orientation in the first principal direction as an example.

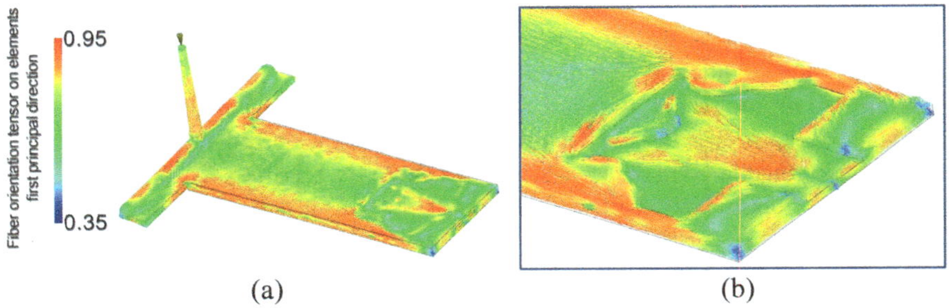

Fig. 6. Fiber orientation tensor first principal direction (a), zoomed in cross-sectional view through the midsection (b).

Mapping is necessary so that these process-specific material properties can be transferred to the FE model of the topology optimization in the *Data Transfer* step. Weighted element mapping of the MpCCI mapper from Fraunhofer SCAI is used for this purpose [11]. Figure 7 shows an example of the mapping of the Young's modulus in the first principal direction from the Moldflow FE mesh to the Abaqus FE mesh.

After the data has been transferred to the mesh of the topology optimization, the Abaqus input files, which contain distributions with element-specific properties, are overwritten with this data in the final step of the interruption *Overwrite distributions and calculate objection function* and the FEA is performed. The density of FRP elements is considered as 1, whereby the fully determined process-specific material properties are used in the analysis. The reaction force at the clamping is calculated, which should be maximized (objective function). Tosca Structure then calculates the density distribution for the next iteration and the interruption starts all over again.

Once the convergence criterion has been met, which requires the objective function and element density delta criterion to be fulfilled, the topology optimization is completed. Finally, a new static analysis is carried out in the *Optimized hybrid* step with the design found to eliminate any existing intermediate densities.

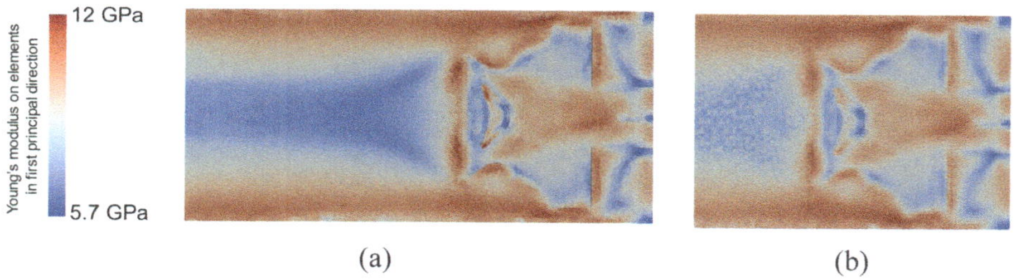

Fig. 7. Cross-sectional view through the midsection of the FE mesh of Moldflow with Young's modulus on elements in first principal direction (a) and cross-sectional view through the midsection of the FE mesh of Abaqus with mapped Young's modulus on elements in first principal direction (b).

3 Numerical Example

In this numerical example, the structure of a flat metal insert is to be optimized which is part of a FRP-metal-hybrid produced with injection molding. The corresponding dimensions and the load case can be seen in Fig. 8.

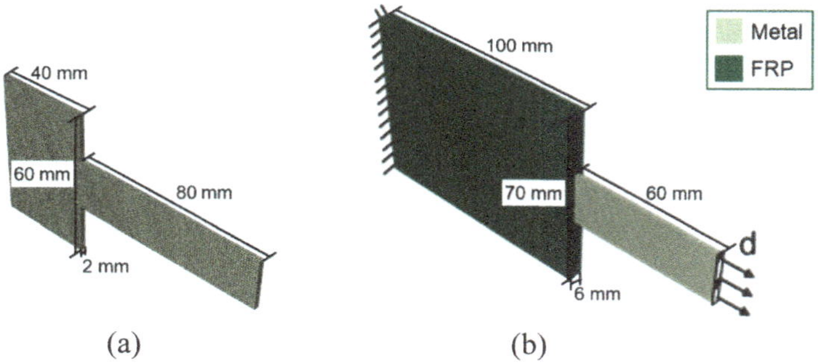

Fig. 8. Metal insert geometry (a), hybrid geometry (b) with the identical load case as in Fig. 4 including the residual stress analysis as first step.

S700 is used for the metal insert with elastic material properties and a Young's modulus of 210 GPa. A3WG6, which is polyamide 6.6 with 30% glass fiber reinforcement, is used for the FRP with elastic-plastic material properties. Those properties are based on the stress-strain curve of the data sheet of A3WG6 and integrated in Abaqus with material calibration [12]. Linear hexahedral elements with an edge length of 1 mm are used. The following contact parameters were used, which can be estimated from pull-out tests performed on the test bench of unoptimized hybrid structures (see Table 1).

Figure 9(a) shows the reaction force-displacement curve of the pull-out test of such an unoptimized hybrid. The displacement starts with a negative value because the hybrid contracts in the first step of the FEA as a consequence of the initial stresses in the FRP after injection molding. At a total pull-out displacement of 0.7 mm a drop of reaction force occurs. After the drop the reaction force increases with further displacement. The

Table 1. Cohesive and damage contact parameters

K_{nn}, K_{ss}, K_{tt}	σ_{max}, τ_{max}	δ_f
40 MPa/mm	2 MPa	0.1 mm

displacement is only depicted until a value of 0.6 mm (total displacement of approx. 1 mm) with a reaction force of 22.6 kN because this work focuses on the premature failure taken place at the mentioned reaction force drop. Analyzing the contact status during the reaction-force drop shows that most of the contact surface detaches there. As most cohesive contacts are not in contact anymore after 1 mm total displacement (see Fig. 9(b)) the still increasing reaction force is due to form fit.

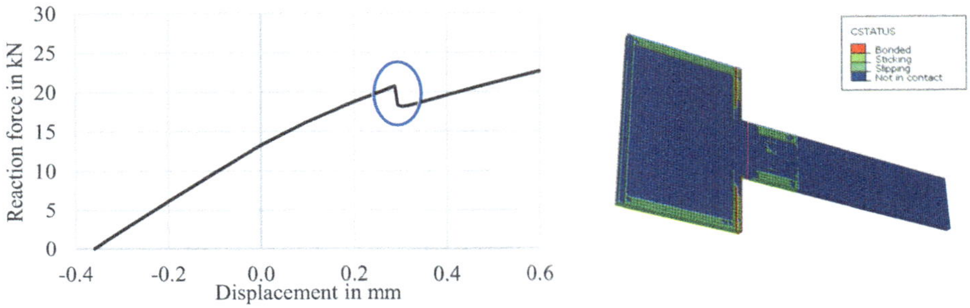

(a) (b)

Fig. 9. Reaction force in relation to the displacement from pull-out tests of unoptimized hybrid structures (a), contact status after 1 mm displacement in the pull-out test (b).

This premature failure of the cohesive zone should be prevented by maximizing the reaction force for a specified displacement of 1 mm. The volume reduction is set to 30% for the following design space inside the metal (see Fig. 10). To achieve a 2.5D geometry

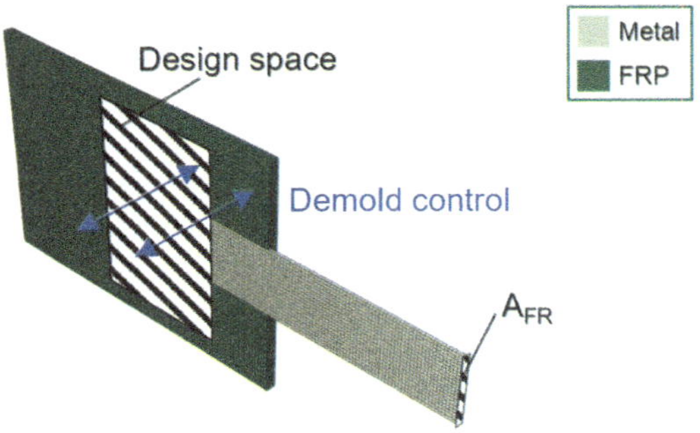

Fig. 10. Cross-sectional view through the midsection of the hybrid with design space, demold control direction and area for the calculation of the reaction force in pull-out direction (A_{FR}).

for the optimized metal insert, so that the FRP can penetrate the resulting cavities during injection molding, a demold control is selected as a manufacturing constraint.

4 Results and Discussion

4.1 Results with the Developed Optimization Method

Applying the method results in a design that has several features. Firstly, it has a support surface on the right-hand side in Fig. 11, which counteracts pull-out. This surface is extended by struts. There is also a compression strut on the left-hand side, which prevents the FRP from contracting too much. An opening can be seen in the middle of the metal insert, which creates a form fit and saves material of the metal insert.

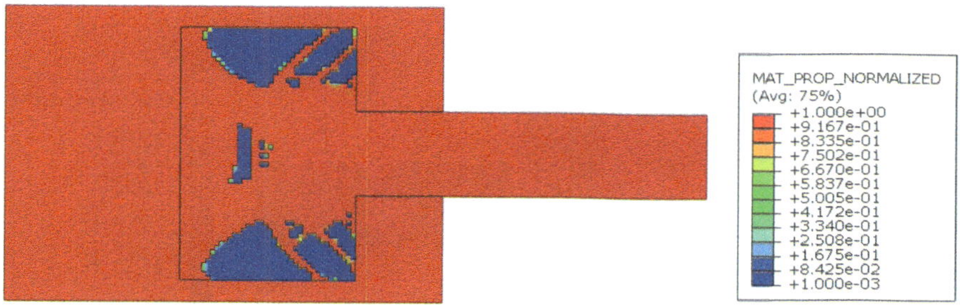

Fig. 11. Cross-sectional view through the midsection of the density distribution by applying the optimization method at convergence iteration.

When evaluating the reaction force-displacement curve in Fig. 12 there is no drop in reaction force and a higher value of 28.3 kN is achieved after 1 mm displacement compared to the unoptimized metal insert. When analyzing the contact status, it can also be seen that the metal insert has only detached slightly and the cohesive contact is still intact in most places, meaning that there is only neglectable premature failure.

4.2 Comparison to Results from a Standard Optimization

A standard topology optimization is carried out where the contact interface and process-specific material properties remain the same as in the initial FE model. In Fig. 13, a kind of "mushroom head" resulted for the optimized design, which only shows a form fit with the original FRP at the boundary of the design space. This can be explained since in the standard optimization the contact interface is not adjusted and remains on the boundary of the design space. Otherwise, there are many intermediate densities still present which indicate a badly converged solution. A cross strut can be seen on the left-hand side in Fig. 13, which also counteracts the contracting FRP as in Fig. 12. However, this strut is not connected to the rest of the metal insert.

The reaction force-displacement curve in Fig. 14(a) is less linear, and a lower value for the reaction force of 27.4 kN is achieved than before with application of the method.

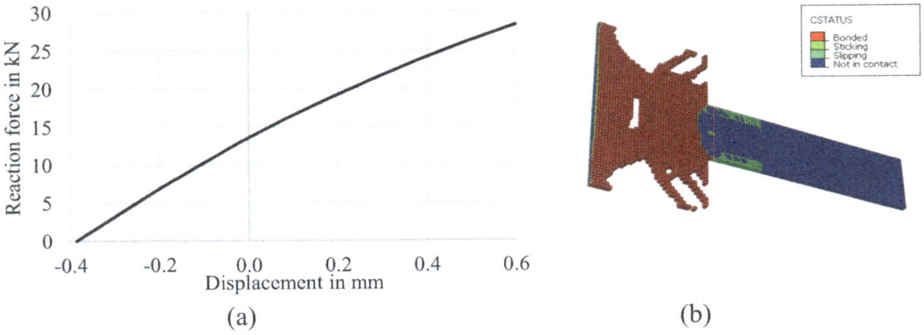

Fig. 12. Reaction force in relation to the displacement from pull-out tests of the optimized hybrid structure by applying the optimization method (a), contact status after 1 mm displacement in the pull-out test (b).

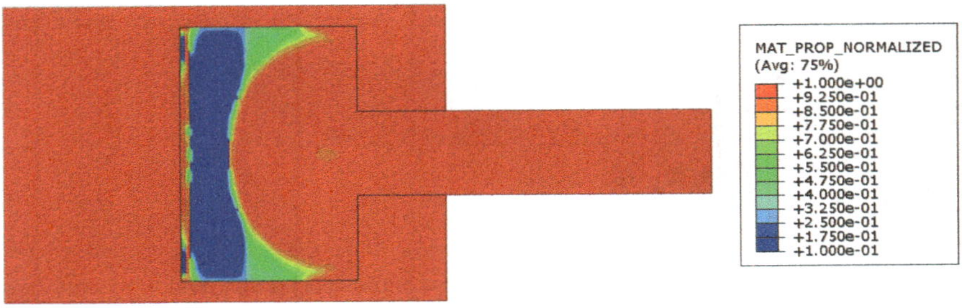

Fig. 13. Cross-sectional view through the midsection of the density distribution with standard optimization at convergence iteration.

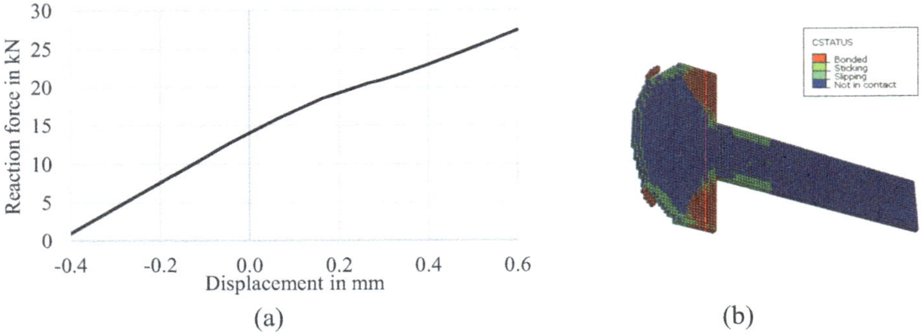

Fig. 14. Reaction force in relation to the displacement from pull-out tests of the optimized hybrid structure with standardized optimization (a), contact status after 1 mm displacement in the pull-out test (b).

Considering the contact status, a large part of the metal insert has detached from the FRP and the cohesive zone and therefore the hybrid component have failed prematurely.

The comparison shows that a 3.3% increased reaction force can be reached by applying the developed optimization method which also results in a stiffer FRP-metal-hybrid. Moreover, the cohesive zone remains mostly intact by applying the method without premature failure of the hybrid in contrast to the standard optimization result.

5 Conclusion

With FRP-metal-hybrids advantageous material properties of both materials can be integrated in one part and when used on the right positions inside the part, weight can be saved. To increase the lightweight potential of such hybrid structures even more, manufacturing related material properties should be considered in the design process. In this paper an optimization method for FRP-metal-hybrids manufactured through over molding of a metal insert with FRP within the injection molding process is presented to support the product developer in the design process of such hybrids.

The optimization method considers process-specific material properties of the injection molding process during a sensitivity-based topology optimization of the metal insert in Tosca Structure. Therefore, in each iteration of the topology optimization an interruption takes place after the density distribution is calculated. The contact pairs between FRP and metal are adjusted according to the density distribution and a mold filling simulation based on the smoothed surface geometry of the interim result of the metal insert is carried out in Moldflow. The resulting fiber orientations, engineering constants and initial stresses of the FRP elements are exported and mapped on the FE mesh which is used during the topology optimization. Those process-specific material properties are consequently considered in the FEA and the interruption is finished. The sensitivities are calculated by Abaqus and the density distribution for the following iteration by Tosca Structure using the MMA. When convergence is reached the topology optimization is finished, resulting in an optimized FRP-metal-hybrid.

A numerical example of the optimization of such a hybrid consisting of S700 and A3WG6 (polyamide 6.6 with 30% glass fiber reinforcement) is shown. For a prescribed displacement of 1 mm and volume reduction of 30% of the design space within S700 the reaction force should be maximized. Geometries for the metal insert resulting from the developed method are different to those resulting from a standard topology optimization. The hybrid structure optimized with the developed method reaches a higher reaction force and premature failure of the cohesive zone is avoided in contrast to the hybrid structure optimized with a standard topology optimization.

To evaluate the influence of the fiber content on the optimized design, in future work polyamide 6.6 with 15% and 50% fiber reinforcement will be examined in the developed method. A different load case will be examined as well. The optimized designs will be produced and tested on a test bench to validate the proposed optimization method.

Acknowledgements. This publication resulted within the project 'HyTop: Development of a coupled topology optimization method for injection-molded, short-fiber reinforced thermoplastic-metal hybrids to support the product developer in design synthesis', which is funded by the German Research Foundation (DFG). The support of the DFG is gratefully acknowledged.

References

1. Zhao, G.: Spritzgegossene, tragende Kunststoff-Metall-Hybridstrukturen: Konstruktion, Prozessanalyse und Charakterisierung, Zugl.: Erlangen-Nürnberg, Univ., Diss., 2001, Lehrstuhl für Kunststofftechnik, Erlangen (2002)
2. Henning, F., Weidenmann K., Bader B.: Hybride Werkstoffverbunde, In: Henning, F., Moeller, E. (Hrsgg.): Handbuch Leichtbau: Methoden, Werkstoffe, Fertigung, Hanser, München, Wien, pp. 415–428 (2011)
3. Bendsøe, M.P., Sigmund, O.: Topology Optimization. Springer, Heidelberg (2004)
4. Albers, A., et al.: Strategische Planung des Entwicklungsrisikos gekoppelter CAE-Methoden. Konstruktion (2022)
5. Shin, D.K., Kim, H.C., Lee, J.J.: Numerical analysis of the damage behavior of an aluminum/CFRP hybrid beam under three point bending. Compos. B Eng. **56**, 397–407 (2014)
6. Paul, H.: Bewertung von langfaserverstärkten Kunststoff-Metall-Hybridverbunden auf der Basis des Verformungs- und Versagensverhaltens, Karlsruhe (2013)
7. Dassault Systèmes 3ds Homepage. https://help.3ds.com/2022/english/dssimulia_established/SIMACAEITNRefMap/simaitn-c-cohesivebehavior.htm?contextscope=all. Accessed 13 Jan 2025
8. Svanberg, K.: The method of moving asymptotes - a new method for structural optimization. Int. J. Numer. Meth. Eng. **24**, 359–373 (1987)
9. Folgar, F., Tucker, C.: Orientation behavior of fibers in concentrated suspensions. J. Reinf. Plast. Compos. **3**(2), 98–119 (1984)
10. Mori, T., Tanaka, K.: Average stress in matrix and average elastic energy of materials with misfitting inclusions. Acta Metall. **21**, 571–574 (1973)
11. Spiess, H., Oeckerath, A., Landvogt, B.: MapLib Manual Version 2013.0, Fraunhofer Institute for Algorithms and Scientific Computing (SCAI): Sankt Augustin (2018)
12. Albis Homepage. https://www.albis.com/de/products/download/doc/de/SI/basf/UltramidA3WG6.pdf. Accessed 13 Jan 2025

Foldable CFRP Structure for High-Power Solar Arrays for Nanosats

Julia Wallner(✉), Linmei Li, Valentin Dalbauer, and Sebastian Kölbl

Deggendorf Institute of Technology – Technology Campus Hutthurm, Hochleiten 1, 94116 Hutthurm, Germany
julia.wallner@th-deg.de

Abstract. A CFRP (carbon fiber reinforced polymer) support structure for a foldable solar sail intended for energy generation in space is being developed. The composite must be no thicker than 0.3 mm and must store sufficient strain energy to enable deployment. The self-deployment function must operate effectively within a temperature range of -60 °C to +120 °C, and the CFRP structure must withstand temperatures ranging from -90 °C to +120 °C. The foldability of the structure requires precise material characterization and accurate distribution of the two matrix materials, which are applied sequentially.

The magnetic multi-matrix (MMM) process was developed to meet these requirements. This innovative manufacturing technique ensures the mechanical integrity and electrical performance of the solar cells. The combination of the process with specially selected materials is crucial to achieve a composite with optimal flexibility and stability. By integrating advanced materials and optimized production processes, an efficient and scalable solution for flexible solar energy applications in space is achieved. This technology can be utilized for many other applications where lightweight construction, multifunctionality, and reliability are critical and where folding or unfolding is advantageous.

Keywords: Multi Matrix · Function Integration · Self-Deployment · Composite

1 Introduction

The use of solar panels in space is particularly important for nanosat applications, as they ensure efficient energy supply. To achieve high energy yield, the solar panels must have a large surface area.

At the same time, transportation costs to space are high, which is why compact and lightweight components are preferred. A proven solution is the foldable solar cell carrier structure, which is realized with hinges and meets both requirements: a large surface area in space that can be folded into a compact volume for transport. Additionally, carbon fiber reinforced polymer (CFRP) structures are widely used in aerospace. These specifics led to the development of foldable and self-deploying CFRP solar panels without hinges.

F. Endress et al. (Eds.): MSLD 2024, *Proceedings of the Munich Symposium on Lightweight Design 2024*, pp. 47–54, 2025.
https://doi.org/10.1007/978-3-032-01360-6_5

2 Material Selection

2.1 Technical Requirements of the Solar Panel for a Nanosat

The selection of materials for the solar panel of nanosats is crucial for its functionality. Different matrix materials are used within the CFRP structure to achieve the desired properties for folding and self-deployment without hinges. The flexible matrix material in the joint areas allows for the folding function, while the rigid areas provide the necessary support for the solar cells. The folding design requires a maximum component thickness of 0.3 mm.

Additionally, the materials must be able to withstand extreme temperatures, ensuring self-deployment at -60 °C and survival at -90 °C. This necessitates extensive material investigations and technological developments to identify the most suitable materials for these challenging conditions.

2.2 Basic Requirement of Material

The component of interest is a foldable solar cell carrier for a nanosat. The carrier structure comprises two distinct areas with different mechanical properties:

- **Non-foldable area**: This area needs to be solid and rigid to secure and support the solar cell foils.
- **Foldable area**: This area must be flexible to ensure foldability. An example of a nanosatellite solar array is shown in Fig. 1, where all the rigid areas (angular areas) are separated from the flexible areas (linear areas).

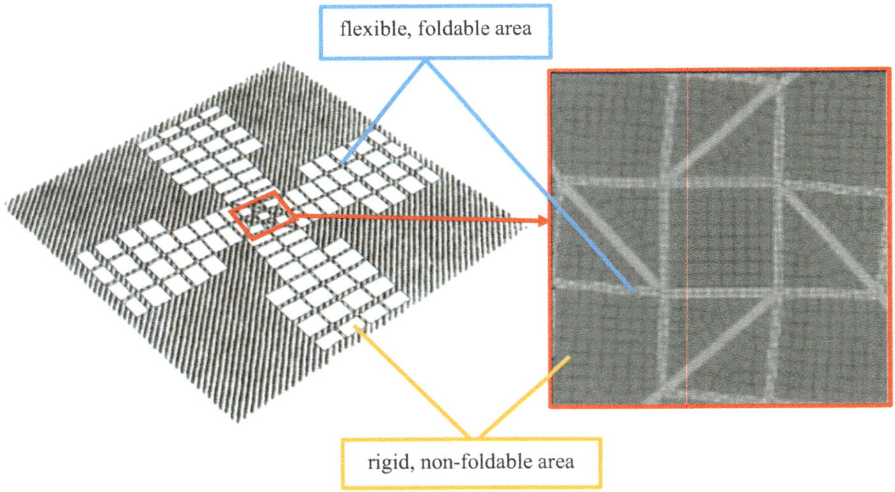

Fig. 1. An overview (left) and a close-up (right) illustration of the foldable solar array for a nanosatellite, constructed from foldable CFRP, featuring both rigid areas (indicated by yellow lines; angular areas) and flexible areas (indicated by blue lines, linear areas)

2.3 Carbon Fiber Fabric and Epoxy Resin

The differing requirements necessitate a combination of materials. Three basic materials were required: a reinforcement structure, a matrix for non-foldable areas, and matrix material for the foldable areas. Materials used in the aerospace industry must be approved for such use. Therefore, the search initially focused on aerospace-approved materials.

In the technical specifications, both thickness and weight are constrained. The carbon fiber fabric is required to have an area weight ranging from 90 to 100 g/m^2 to comply with the specified weight and thickness criteria.

The properties of the non-foldable area were easily achieved with space-suitable epoxy resin and carbon fiber fabric.

2.4 Silicone

Carbon fiber fabric is foldable. However, it cannot self-deploy or unfold on its own. Once impregnated with epoxy resin and cured, the fabric becomes rigid and non-foldable.

To achieve self-deployment, the carbon fiber fabric requires the support of matrix materials. An elastic material matrix system is proposed for this purpose.

Furthermore, due to the requirement of -60 °C for functionality and -90 °C for survival temperature, an elastomer to consider is silicone rubber. Silicones exhibit high thermal stability for plastics, up to approximately 250 °C, and low glass transition temperatures down to -120 °C, which is due to the very strong Si-O bond [1]. This thermic property meets the requirements well. Silicone rubber is often used in the aerospace sector to protect electronic components, so aerospace-approved silicone rubber can easily be found.

Several companies offer silicone rubber for aerospace applications, including Momentive, Wacker Chemie, Dow Chemical, Avantor Sciences and others. The silicone rubbers were initially identified and subsequently evaluated based on their technical properties, such as glass transition temperature (T_g) and viscosity, as well as cost and delivery time. Based on these evaluations, NuSil CV14-2500 was selected for further testing.

Differential Scanning Calorimetry (DSC) was employed based on DIN EN ISO 11357-5 to determine the curing range. Figure 2 shows the DSC curve between 0 and 150 °C of NuSil CV14-2500 during heating in nitrogen atmosphere with a heating rate of 3 K/min.

The peak analysis shows an onset temperature of 55 °C and a peak temperature of 75 °C. Therefore, the curing temperature was set to 65 °C to minimize internal stresses and prolonged reaction times.

The T_g of the fully cured silicone rubber was determined using DSC, as T_g is a crucial requirement for the self-unfolding function. The sample was measured over two temperature cycles, heating and cooling at a rate of 20 K/min. The DSC curve, shown in Fig. 3, indicates that the selected silicone rubber has a T_g of around -120 °C, meeting the criteria for the self-unfolding function.

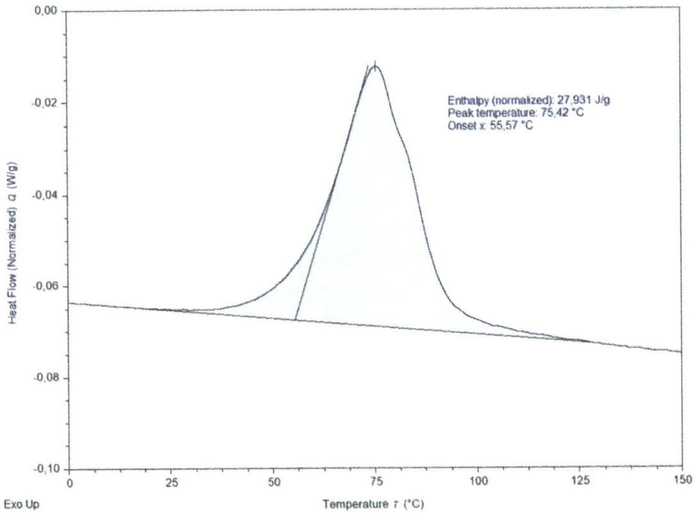

Fig. 2. Section of a heating curve of a DSC-measurement from 0 °C to 150 °C to determine the curing temperature of NuSil CV14-2500; the heating rate was set to 3 K/min

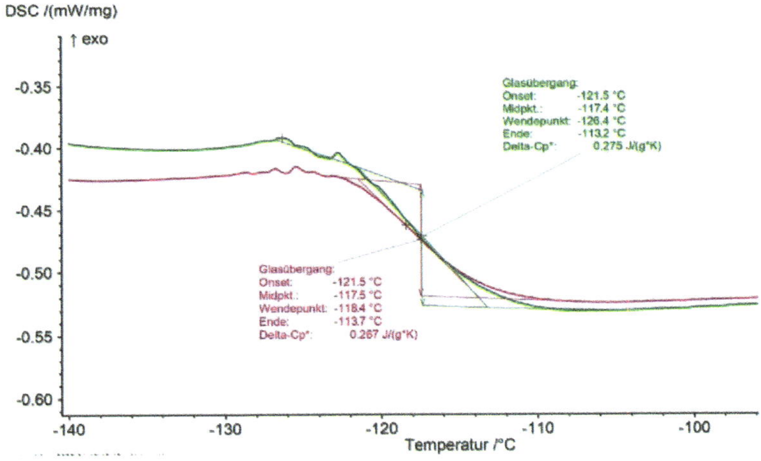

Fig. 3. Section of two heating curves with 20 K/min heating rate of DSC-measurements from −140 °C to −96 °C to determine T_g of fully cured NuSil CV14-2500; T_g was determined using the inflection method

2.5 Silicone Contamination Test

The manufacturer indicates that silicone rubber can be easily contaminated by various substances. To prevent the entire prototype from being contaminated, the materials used in the manufacturing process were tested first. DSC measurements with contaminated samples are conducted to determine the total curing enthalpy. The curing process was carried out according to the production process. The material was contaminated during mixing. As a reference enthalpy, the curing enthalpy of pure, not contaminated silicone was determined to be 28.3 J/g.

Table 1. Curing enthalpy and calculated degree of cure of pure and differently contaminated silicone samples

Material	Curing Enthalpy	Degree of Cure
Pure material	28.3 J/g	100%
Pig hair brush	22.1 J/g	78.1%
Wood	27.8 J/g	98.2%
PE film	19.8 J/g	70.0%
Carbon fiber	24.3 J/g	85.6%

Table 1 shows the determined curing enthalpy and the calculated degree of cure of differently contaminated silicone. Perforated films and brushes are not used during processing. The sizing of the used carbon fibers hinders the curing of the silicone rubber. The search for suitable sizing is part of ongoing research. However, a curing rate of 85.6% was acceptable for a prototype.

3 Methods

Within this research, three technical methods to produce CFRP were considered:

- Resin Transfer Molding (RTM)
- Vacuum Infusion
- Hand Lay-Up

These methods were chosen to determine the best manufacturing process for the specific requirements.

3.1 Resin Transfer Molding

The RTM process represents a promising approach to produce composite materials [2]. In the dissertation of J. W. Krollmann [3], the possibility of incorporating multiple matrix systems using the RTM process was investigated. A specific approach involved isolating the boundary areas by using silicone seals.

To verify the feasibility of this process for complex geometries, a section of a solar cell carrier structure was used. Silicone cords were employed to isolate the boundary areas and enable controlled resin injection. Figure 4 shows the RTM tool of a section (left) and the first part (right), which exhibited significant overflow of silicone rubber in the rigid areas.

The design proved to be extremely challenging, as the process requires a minimal number of injection points while also considering air outlet points. This is particularly demanding for complex geometries. The precise placement of the silicone cords and ensuring an even distribution of the resin are critical.

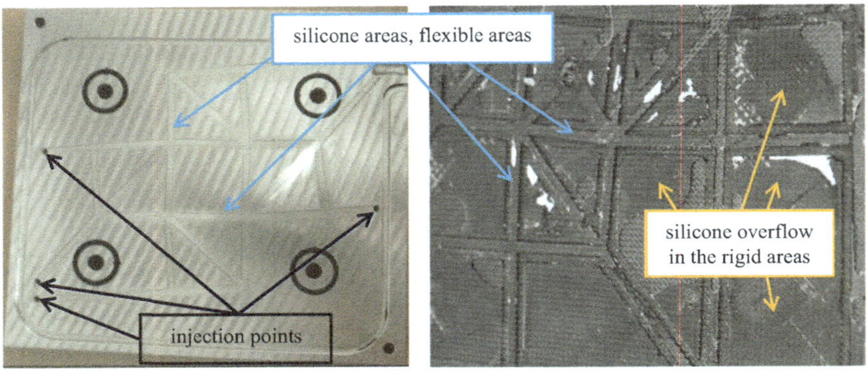

Fig. 4. The RTM tool for a part of the solar panel of a nanosatellite (left) and the first test part (right) include several key features. The silicone areas or flexible areas are indicated by blue arrows, while the injection points are marked by black arrows. The silicone overflow into the rigid areas is highlighted by yellow arrows. Additionally, the white spots on the right are the foamed silicone that occurred during processing under vacuum

The manufacturing process was very demanding, as parameters such as temperature, pressure, and viscosity had to be carefully adjusted to ensure optimal resin injection. Despite intensive adjustments, it was difficult to achieve satisfactory results. Due to the high complexity and challenging parameter control, the process was ultimately deemed unsuitable and was discarded.

3.2 Vacuum Infusion

Another possible manufacturing method is vacuum infusion with multiple matrix systems. This method is based on a standard vacuum infusion process. The modification involves the simultaneous introduction of several matrix materials. A large transition zone between the materials is tolerated [4].

In a further developed version of the process, a porous material is placed on top of the fiber material. This is used to stop the matrix material that reaches the transition zone first. A rough positioning of the transition zone can thus be realized with this variant of the vacuum infusion technique [5].

However, similar to 3.1, these two variants also require one resin inlet and outlet per area of matrix material, which makes the set-up extremely complex. Due to the lack of separation, no complex geometry is feasible, only a very imprecisely defined transition positioning. This method is not suitable for the satellite carrier structure, which requires precise folding. In addition, the combination of silicone and epoxy resin cannot be cured simultaneously. Both variants are not suitable for series production.

No tests were therefore carried out with these methods and the processes were excluded.

3.3 Hand Lay-Up

The oldest, yet still proven simple, and adaptable method for manufacturing fiber-reinforced components is the hand lay-up process [2]. The first prototype was produced using this method. Since the epoxy resin inhibits the curing of the silicone rubber, the silicone was applied first. The epoxy areas were covered to prevent the silicone from overflowing.

Advantages:

- The choice of materials is less critical compared to other methods, such as for RTM.
- High flexibility and applicability for various geometries

Disadvantages:

- The quality of the component heavily depends on the craftsmanship of the worker.
- Not suitable for mass production

All three methods exhibit specific weaknesses in the production of CFRP multi-matrix components. Through the results and the test of the methods RTM and hand lay-up, an automated hand lay-up process, referred to as the "magnetic multi-matrix (MMM) process", was tested.

The MMM process replaces manual labor with the use of robots. In modern industry, robots are widely used in numerous areas, particularly in mass production.

The MMM process is characterized by a wide selection of materials, flexibility in component geometry, and ease of handling, thereby eliminating the previous disadvantages of conventional hand lay-up methods. It enables precise matrix distribution, which is applied sequentially. After the production of the initial prototypes, the models intended for space missions will also be produced using the MMM process.

4 Discussion and Outlook

This work has led to the development of a series process for self-deployable CFRP structures. This comprehensive development encompasses materials, methods, and process automation. The development enables the mass production of multi-matrix CFRP components. However, further development is still required, which focuses on material improvement and process optimization.

The material improvement is initially targeted at the curing and contamination of silicone rubber. According to the manufacturer, the sizing of the carbon fibers consists of an epoxy resin solution that inhibits the curing of the silicone rubber. Various approaches to solving this problem include washing off the sizing with chemicals, burning it off, or coating with alternative sizings. All possible methods will be explored in material improvement to enhance the curing degree of the silicone rubber.

Process optimization aims to enhance the degree of automation in covering the epoxy area by utilizing robotics and advanced techniques such as AI.

The process, which has been submitted for a European patent, is intended to be applied to other solar cell structures in the future. The ease of scalability and adaptability to different geometries are decisive advantages.

References

1. Kaiser, G., Schmölzer, S.: Handbuch DSC, Netzsch Gerätebau GmbH, Netsch, Selb (2015)
2. Flemming, M., Ziegmann, G., Roth, S.: Fasern Verbundbauweisen: Fertigungsverfahren mit duroplastischer Matrix, 1st edn. Springer, Heidelberg (1999)
3. Krollmann, J.W.: A contribution to integral fiber reinforced hinges based on carbon fiber reinforced elastomers and the hybrid-matrix approach, doctoral dissertation, pp. 31–57. Technical University Munich, Munich (2017)
4. Fang, X., Lin, W., Xie, M., Finn, S.R.: Multi-Resin Composite Articles and System and Method of Manufacture, US 2008/0305340 A1 (2008)
5. Krollmann, J., Snajdr, R., Paz, M., Zaremba, S.; Drechser, K.: Hybrid-matrix approach: How to overcome the conflict of matrix selection? In: Proceedings of the Regional Conference Graz 2015 - Polymer Polymer Processing Society PPS: Conference Papers, Graz (2016)

Functionalization of Endless Glass Fiber-Reinforced Polymers Through Incorporating FeSi Particles

Dominik Santl[1(✉)], Josef Ganslmaier[2], Erik Buchenau[3], Matthias Bleckmann[4], and Philipp Höfer[2]

[1] Airbus Defence and Space GmbH, 85077 Manching, Germany
[2] Institute of Lightweight Engineering, University of the Bundeswehr Munich, 85577 Neubiberg, Germany
[3] Microwaves and Radar Institute, German Aerospace Center, 82234 Oberpfaffenhofen, Germany
[4] Bundeswehr Research Institute of Materials, Fuels and Lubricants (WIWeB), 85435 Erding, Germany

Abstract. To functionalize and adjust certain material properties the addition of micro- or nanoparticles to composite materials is a well-known method. Various filler materials and particle sizes can be used and added to different matrix materials to improve for instance mechanical, thermal, electrical or electromagnetic properties. This study investigates failure mechanisms and the short beam strength of ferrosilicon-filled endless glass fiber-reinforced polymers. The fundamental influences are investigated through the manufacturing of a sample series with three particle types and five different particle contents. The material composition in each specimen is analyzed using density measurements and the fiber to filler ratio. The short beam strength is determined according to DIN EN 2377 and DIN EN ISO 14130.

Keywords: functionalized fiber composites · short beam strength · filler incorporation

1 Introduction

The combination of materials is becoming increasingly important as the demand for high performance materials and components grows. The focus is not only on performance, but also on durability and reliability. Materials functionalization is a broad area of research that deals with the integration of additional functions and properties into existing materials. Methods of adding functionality range from modifying molecules, to changing material composition, to adjusting surface morphology. Depending on the application, examples include improving adhesion by introducing additional chemical groups [1,2] or changing the surface morphology [3,4], modifying physical properties such as thermal conductivity [5,6], or improving the aging behavior of a material [7]. Properties

F. Endress et al. (Eds.): MSLD 2024, *Proceedings of the Munich Symposium on Lightweight Design 2024*, pp. 55–67, 2025.
https://doi.org/10.1007/978-3-032-01360-6_6

are often adjusted by adding fillers in the form of particles of different sizes, ranging from nanometers to micrometers, to a base material. This is particularly common in the manufacture of materials for electromagnetic absorption. Dielectric fillers such as carbon black, carbon nanotubes [8] and silicon carbide or magnetic materials, usually iron alloys [9], are introduced and the electromagnetic waves are absorbed by loss mechanisms in the fillers [10]. State of the art is the use of polymer matrices in which the particles are embedded. This ensures uniform and easy incorporation of the fillers. However, the disadvantage is that the mechanical properties of the polymers are further reduced by the addition of fillers, depending on the volume content and size [11]. In order to achieve higher mechanical properties, the incorporation of particles into endless fiber-reinforced composites is considered. As a base material, they offer high mechanical strength and are already a well established material. Functionalization is also very widespread in the field of fiber reinforced polymers. Primarily the focus here is on increasing the mechanical properties, adapting matrix systems and surface structures. For example, Bard S. [12] investigated the thermal conductivity by incorporating carbon black particles into the epoxy resin, whereas Rajaei M. et al. [13] changed the mechanical properties and flame retardancy by integrating ammonium polyphosphate and talc.
In this work the incorporation of ferromagnetic particles into endless glass fiber reinforced polymers (GFRP) to modify the electromagnetic properties, in particular to integrate the function of absorbing electromagnetic waves, is studied. The state of the art process to incorporate fillers into endless fiber-reinforced polymers is through the integration of particles in the resin system. Nordlund M. et al. [14] researched, that this procedure leads to a disadvantages, uneven distribution of the particles, because of a filtration effects through the closely packed fibers. Using a novel process, in this work, the fillers are first introduced into a fiber bundle. Afterwards the particle-filled roving can be further processed into any semi-finished product or fiber composite.

This study focuses on the influence of different filler types on the mechanical performance. A sample series with three particle types and particle contents of a ferrosilicon alloy, ranging from 5 vol.% to the maximal achieved vol. content, roughly 35 vol.% is manufactured and analyzed. In particular the effect on the fiber to resin and particle to resin adhesion in a short beam shear test according DIN EN 2377 and DIN EN 14130 is investigated. Alongside an analysis of the changing material composition as a result of the particle incorporation, the failure modes of the specimen, the fracture surfaces and the mechanical results in terms of short beam strength and flexural modulus are analyzed.

2 Experimental

2.1 Materials

For the experiments, FeSi alloy powders with two particle size distributions with spherical shape from Höganäs AB (Höganäs, Sweden) were used. Ferrosillicon is a common alloy used for electromagnetic wave absorption with high permeability and good absorption properties [15]. The wet chemical analysis showed 3.5 wt.% Si and balance Fe. The particle size distributions shown in Fig. 1 were determined using the CAMSIZER X2 optical particle analyzer from Microtrac Retsch GmbH (Haan, Germany), which measured a mean particle size of D50 = 15 μm and D50 = 38 μm, respectively. To test a variation in particle shape, the D50 = 15 μm powder was ball milled to produce platelet-shaped particles. The powder was milled in ethanol with 10 mm diameter tungsten carbide balls at a mixing ratio of 1:1:5 for 2 h at 500 rpm. The resulting shape and particle size distribution was observed using microscope images and the particle analyzer (Fig. 1c). The S2 glass roving PD PowerStrand HMC9-340 T5(68) 340tex from PD-Glasseiden (Oschatz, Germany) was used as the fiber material because of its high strength, good availability and low dielectric constant compared to E-glass. The resin system used was Hexflow® RTM 6 epoxy from Hexcel Cooperation (Standford, USA), which is ideal for high particle filled composites due to its low viscosity specifically designed for resin infusion, high glass transition temperature, and high strength [16].

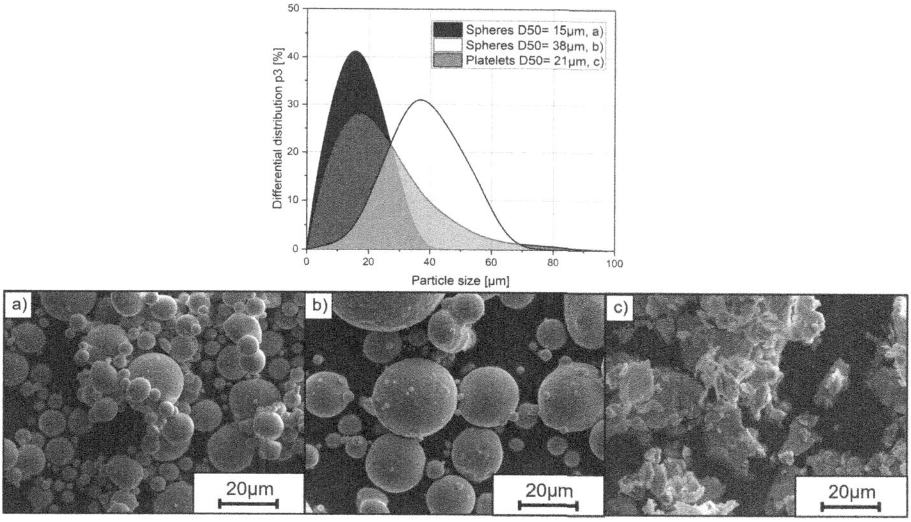

Fig. 1. Measured particle size distributions of the powders and scanning electron microscope (SEM) images of the particles; a) spherical, D50 = 15 μm; b) spherical, D50 = 38 μm; c) platelet shaped, D50 = 21 μm; FOV = 100 μm; 1 kV;

2.2 Composite Fabrication and Sample Preparation

The manufacturing process of the GFRP specimen is shown in Fig. 2 and consists of several steps. The first step is the incorporation of the particles into the endless

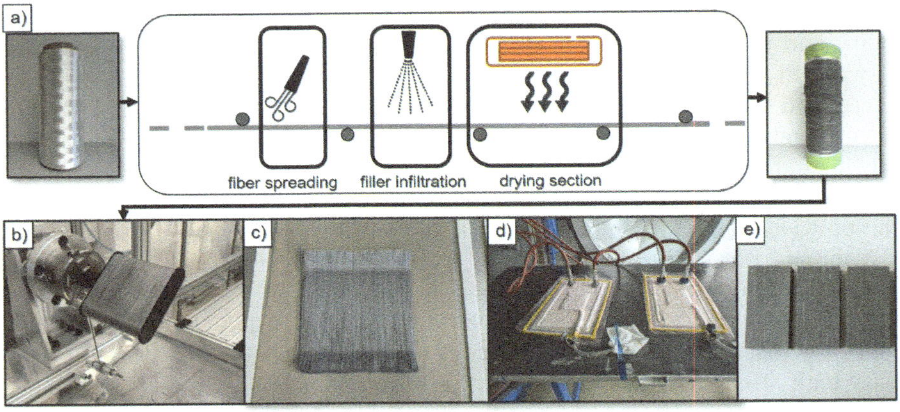

	Targeted filler content [vol.%]				
FeSi, spherical, D50 = 15 μm	10	15	20	25	35
FeSi, spherical, D50 = 38 μm	10	15	20	25	35
FeSi, platelet shaped, D50 = 21 μm	5	7.5	10		

Fig. 2. Schematic of the filler incorporation process and the composite sample manufacturing with the targeted filler contents

glass fiber roving, as shown in the schematic Fig. 2a). The starting material, the glass fiber roving, is fed through the process. In the first step, the roving filaments are spread and a dispersion containing the filler particles is applied. By varying the filler content in the dispersion, the resulting filler volume in the glass fiber can be tailored. The solvent in the dispersion is then evaporated in a drying section using infrared heaters, and at the end of the process the roving is wound back onto a spool. The filled roving is then used to produce the dry preform by winding it onto a winding core (Fig. 2b). The core is weighed before and after winding in order to determine the amount of filler in the roving based on the length of the roving and the weight of the preform. After this step, the preform is separated from the core (Fig. 2c) and placed in a vacuum infusion assembly in an unidirectional layup. The preform is then infiltrated with the epoxy resin and cured for 2 h at 180 °C at 6 bar in an autoclave (Fig. 2d). The resulting test plate is then milled on the top and bottom surface to 2 mm or 3 mm depending on the test standard, cut into test specimens using a diamond circular saw and finally ground to a length of 20 mm and width of 10 mm (Fig. 2e). A total of 13 material configurations were produced and tested. For both types of spherical filler, the filler content was targeted in 5 vol.% steps from 10 vol.% to the maximum achievable filler content of 35 vol.%. For the platelets, the maximum achievable content was 10 vol.% filler and therefore three configurations with 2.5 vol.% steps were used.

2.3 Analysis of the Material Composition

The introduction of fillers into a fiber-reinforced polymer turns the two-component system into a three-component system, making the determination of the material composition much more difficult. Classical methods such as acid digestion, resin burn off, and optical microscopy based techniques are not applicable or are inadequate for these materials. For this reason, a method has been developed that allows the determination of each component. As mentioned above, the filler content in the roving is already determined during the production of the preform. This provides the ratio of fiber to filler volume. After the fiber composite has been produced, the density of the samples is determined in a Ultrapyc 5000 gas pycnometer from Anton Paar Germany GmbH (Ostfildern-Scharnhausen, Germany). The density of each component, the total density of the GFRP, and the fiber/filler volume ratio are used to accurately determine the material composition.

2.4 Mechanical Testing

The material configurations were tested in a short beam shear test. This test subjects a beam to bending in a three point bending fixture. With a high thickness to length ratio, the objective is to minimize the tensile and compressive stresses and maximize the shear stress. Therefore, this test provides information on the quality of the interface adhesion between the fibers and fillers to the resin system. The test standards DIN EN 2377 and DIN EN ISO 14130 were applied. The dimensions of all samples are 20 mm in length and 10 mm in width and depending on the testing standard, 3 mm (DIN EN 2377) and 2 mm (DIN EN ISO 14130) in thickness. Five samples were tested for each material configuration. Testing was performed on the tensile testing machine Z020 equipped with a 10kN load cell by Zwick und Röll GmbH & Co. KG (Ulm, Germany). During the test, the fracture behavior was recorded with the microscope VH-Z50T and the lens OP-88851 from Keyence Deutschland GmbH (Neu-Isenburg, Germany).

3 Results and Discussion

The results of the sample manufacturing and mechanical testing are analyzed below. This includes the material composition, the failure types, the fracture surfaces and the mechanical characteristics. The following references to the filler content in Sects. 3.2 and 3.3 always refer to the targeted filler volume. As the results of the specimens tested with DIN EN 2377 and DIN EN ISO 14130 testing standards barely differ, the results are shown below partially and representative for both testing standards.

3.1 Material Composition

The measured material compositions are listed below for each filler type in Figs. 3, 4, 5. Starting with the small spherical particles, a large increase in the resin volume content can be seen even with a small amount of filler. By introducing the particles into the GFRP, as can be seen in the micrograph Fig. 3a), the distance between the single glass fiber filaments is increased, thereby a reduction of the packing density of the fibers and an increasing resin volume content can be seen compared to the reference material without filler. Even with very few particles, the resin volume content reaches its maximum, as the particles are placed evenly and at different material depths between the fibers, thus maximizing the gaps that need to be filled with resin. As the filler content continues to increase, the resin content is not changing, as the fiber content only decreases with the increasing filler content. From 20 vol.% filler the behavior changes. Here the resin content begins to decrease, whereas the fiber content decreases only slightly. The micrographs c) and f) show, that the gaps are now more and more filled with particles and effectively reduce the resin content.

Targeted filler content	0 vol.%	10 vol.%	15 vol.%	20 vol.%	25 vol.%	35 vol.%
Filler [vol.%]	0	6.7	13.1	18.1	24.4	35.4
Fiber [vol.%]	52.7	29.9	23.8	18.6	17.1	11.6
Resin [vol.%]	47.3	63.4	63.1	63.4	58.5	53

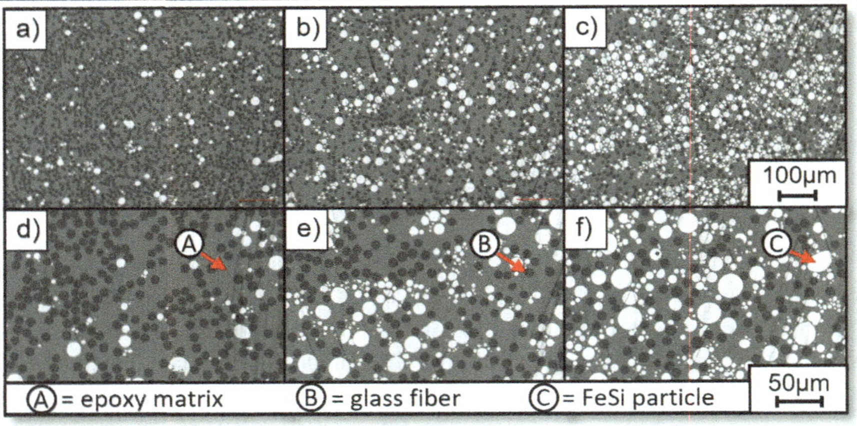

Fig. 3. Material composition and cross-section of the GFRP samples with incorporated spherical FeSi particles (D50 = 15 µm); a), d) 10 vol.%; b), e) 20 vol.%; c), f) 35 vol.%; a)–c) FOV = 750 µm; d)–f) FOV = 300 µm;

The same behavior can be observed with the larger spherical particles. The introduction of few particles increases the resin content in the composite material compared to the reference without filler. In contrast to the smaller particles a more uneven distribution of the fibers can be observed. Due to the larger particle size, the particles are no longer placed evenly between the fiber filaments but divide the fibers into small groups, see Fig. 4a) and b). As the filler content increases, the resin content does not decrease until approx. 20 vol.% filler and is balanced out again by a reduction in the fiber content. The resin content then drops again to a minimum of 55% by volume with increasing particles. Compared to the smaller spherical particles, the fiber content drops to 6.4 vol.% and the resin content does not reach the minimum of the smaller particles. This is due to the larger particles and the corresponding lower packing density. Similar to the other filler types, an increase in the resin content in the composite can be seen through the incorporation of the platelet shaped particles. The micrographs a), b), c) in Fig. 5 show evenly distributed fiber filaments and an alignment of the platelets along the fibers and mainly perpendicular to the sample thickness. As the filler content increases, the resin volume content in the material increases up to a maximum of 66.6 vol.%. In the micrographs e) and f) it can be seen that the platelets agglomerate and additionally create large gaps between the fiber filaments due to their thin shape and large dimensions, which are then filled with resin and thus explain the high resin volume content. This also effectively reduces the fiber volume content to a minimum of 23.6 vol.%.

Targeted filler content	0 vol.%	10 vol.%	15 vol.%	20 vol.%	25 vol.%	35 vol.%
Filler [vol.%]	0	8.8	14.1	20.0	24.3	38.5
Fiber [vol.%]	52.7	30.1	25.5	19	16.4	6.4
Resin [vol.%]	47.3	61.1	60.5	61	59.3	55.2

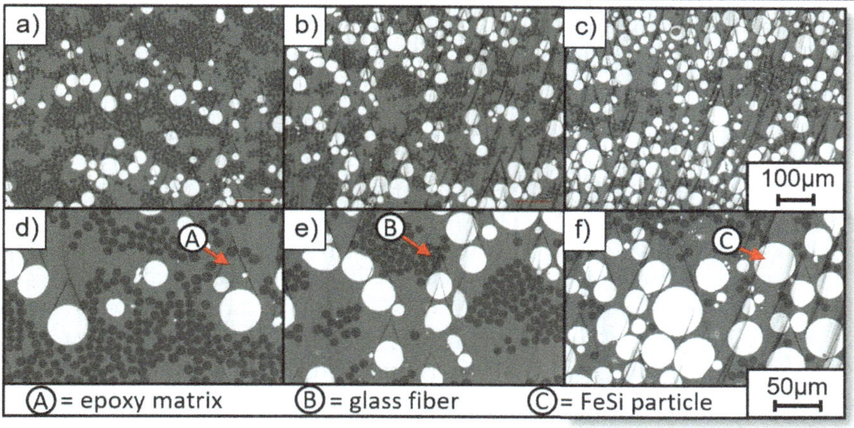

Fig. 4. Material composition and cross-section of the GFRP samples with incorporated spherical FeSi particles (D50 = 38 µm); a), d) 10 vol.%; b), e) 20 vol.%; c), f) 35 vol.%; a)–c) FOV = 750 µm; d)–f) FOV = 300 µm;

Targeted filler content	0 vol.%	5 vol.%	7.5 vol.%	10 vol.%
Filler [vol.%]	0	3.7	7.1	9.8
Fiber [vol.%]	52.7	37.1	30.8	23.6
Resin [vol.%]	47.3	59.2	62.1	66.6

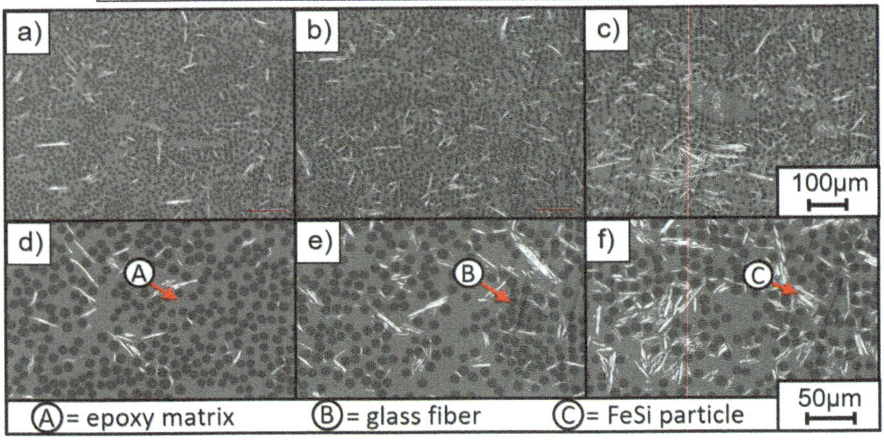

Fig. 5. Material composition and cross-section of the GFRP samples with incorporated platelet shaped FeSi particles (D50 = 21 μm); a), d) 5 vol.%; b), e) 7.5 vol.%; c), f) 10 vol.%; a)–c) FOV = 750 μm; d)–f) FOV = 300 μm;

3.2 Fracture Type

The failure mode of the short beam specimens in the three-point bending test was observed using a Keyence microscope. Figure 6 shows the failure modes of all material configurations using a representative specimen of each. Figure 6a)–e) show the material configurations of the small spherical particles with a D50 = 15 μm, f)–j) the spherical particles with a D50 = 38 μm, k)–m) the platelet shaped particles with a D50 = 21 μm and n) the reference specimen without filler. Interlaminar failure is observed in all figures marked with a circle. The reference, all platelet filler, and the low volume spherical filler specimens failed due to interlaminar shear. It is noticeable that the larger spherical particles weaken the material more than the smaller spherical particles, as interlaminar fractures occur with these up to 20 vol.% filler. In addition, most of the samples with interlaminar fractures also show damage at the bottom of the specimen due to tensile stress. Figures b), c), d), e), i), and j) show pure tensile failure. This type of failure occurs primarily due to the reduced fiber content. It is emphasized that interlaminar failure still occurs in specimens g) and h) despite the reduced fiber content. This confirms the above statement that matrix weakening occurs due to the larger particles and that interlaminar failure dominates up to 20 vol.%.
To examine the fracture surfaces, the samples were broken apart after the test at the resulting crack. One sample of each three filler types with an interlaminar shear failure is shown. The fracture surfaces of the configurations a), f) and k) in Fig. 6 can be seen in the Fig. 7. In a), b), d) and e) the fracture surface of the

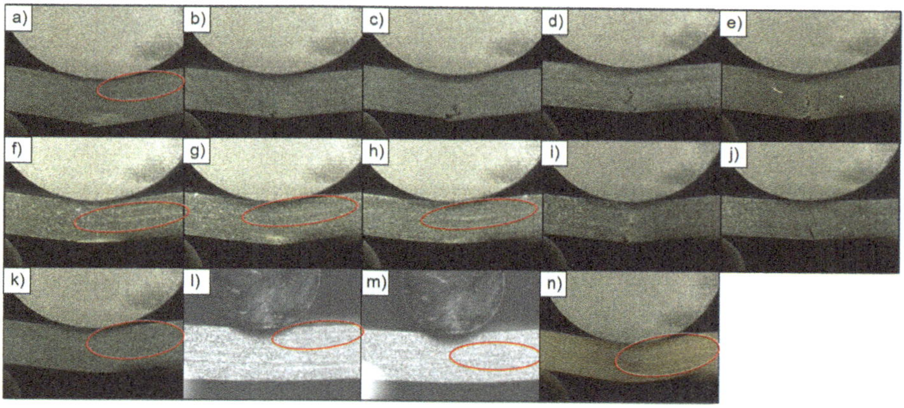

Fig. 6. Micrographs of the short beam specimen showing the fracture type; a)–e) spherical, D50 = 15 μm; f)–j) spherical, D50 = 38 μm; k)–m) platelet shaped, D50 = 21 μm; n) reference without filler; FOV = 8 mm;

spherical particles are displayed, which show a pure failure on the interface adhesion to the polymer matrix. No matrix residues can be observed on the particles. The fibers also show only a few polymer residues, when interlaminar shear failure occurs. It is noticeable that the fiber orientation is influenced locally by the larger spherical particles. This slight undulation explains the stronger material weakening and the interlaminar failure up to 20 vol.% on this material configuration. The fracture surface of the material with the platelet shaped particles is fundamentally different. A rougher surface with multiple small cracks can be observed here. Individual particles are difficult to distinguish from the polymer matrix. The fracture surface indicates that the thin, sharp-edged particles create many small cracks when the material is sheared off and therefore act as crack initiators to a certain extent.

3.3 Mechanical Results

The results shown in Fig. 8 correspond to the specimens with the test standard DIN EN 2377. The mean short beam strength of the reference configuration is 87 MPa, which corresponds to typical values for unidirectional GFRP samples. The incorporation of particles up to 10 vol.% results in a reduction of the short beam strength, which can also be seen as the interlaminar shear strength, to 81 MPa and thus a slight weakening of the fiber and filler adhesion to the resin system. By increasing the particle content up to 20 vol.%, the samples with large spherical particles still fail by interlaminar shear, but the short beam strength is already significantly reduced by 25%. In contrast, the spherical particles with D50 = 15 μm also show a significant reduction in short beam strength due to tensile failure caused by the reduced fiber content. For both spherical configurations, the strength drops to a minimum of 49 MPa and 29 MPa, respectively,

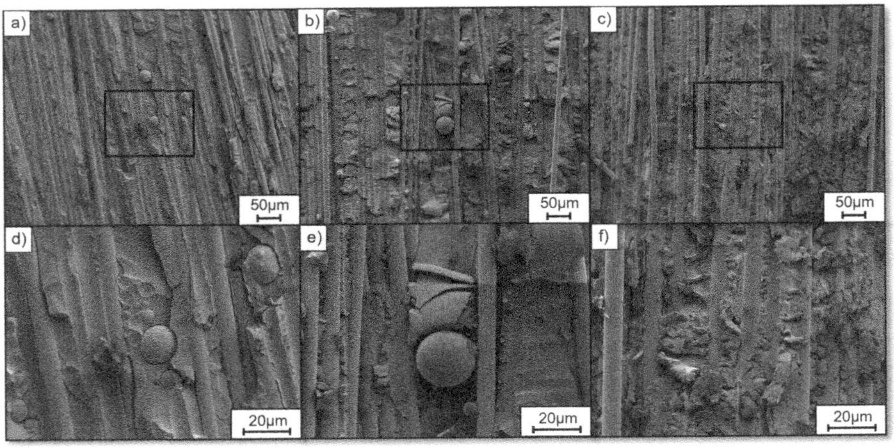

Fig. 7. SEM images showing the fracture surface of the specimen with a), d) 10vol.%, spherical, D50 = 15 µm; b), e) 10 vol.%, spherical, D50 = 38 μm; c), f) 7,5 vol.%, platelet shaped, D50 = 21 µm; a)–c) FoV 500 µm; d)–f) FoV 150 µm; 1 kV;

depending on the fiber content remaining in the configuration with the maximum amount of filler incorporated. The material configuration with the platelet shaped particles shows a different behavior. A reduction in the interlaminar shear strength can also be seen with the addition of only a few particles in the material, but it is noticeable that at 7.5 vol.% and 10 vol.% the shear strength is significantly reduced to approximately 55 MPa. In addition, there is considerable scatter in the results among the specimens of these configurations. From this and the fracture surfaces shown above, it can be concluded that the platelets significantly weaken the material by creating excessive stress concentrations at the particle tips, leading to debonding. With low filler content, the particles act as crack initiators. Depending on the particle orientation and the presence of larger agglomerates, the failure load varies, which explains the high scatter.
In addition to the short beam strength, the flexural modulus was determined between $0.1{\cdot}F_{max}$ and $0.5{\cdot}F_{max}$ according to DIN EN 2562, see Fig. 9. The flexural modulus is defined as a measure of stiffness during the initial bending process. As discussed in Sect. 3.1, the fiber content is significantly reduced by the introduction of particles. In the configurations with the spherical filler, the flexural modulus is reduced by 24% with only 10 vol.% filler. Thereafter, a uniform reduction in stiffness is observed with increasing particle content, especially for the larger spherical particles. It is also noticeable that the values for the larger spherical particles are slightly higher than for the smaller ones. The platelet shaped particles show a different behavior. Here, the configuration with only 5 vol.% filler shows no decrease in stiffness despite the reduced fiber content and the high resin content. The other two configurations with higher filler content also show only a slight decrease in flexural modulus, but significantly higher val-

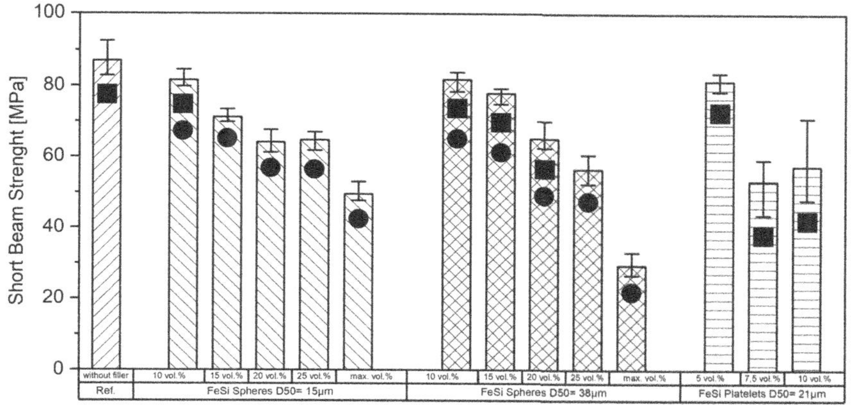

Fig. 8. Short beam strength of all material configurations according DIN EN 2377; ■ = interlaminar shear failure (where measured); • = tensile failure (where measured);

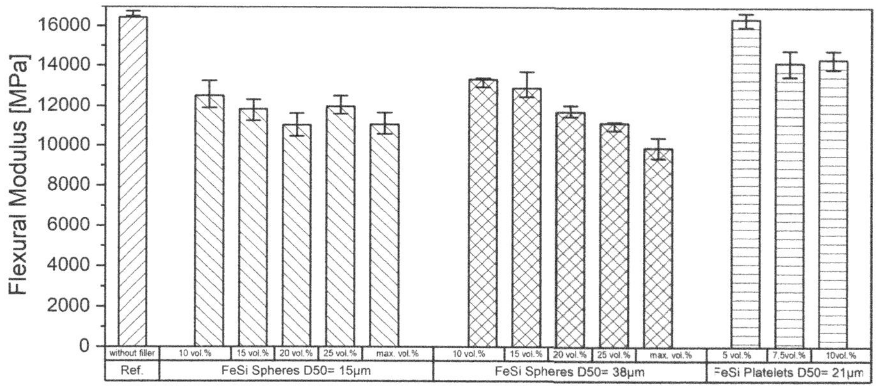

Fig. 9. Flexural modulus derived from the short beam shear test on the basis of DIN EN 2562 determined between $0.1{\cdot}F_{max}$ to $0.5{\cdot}F_{max}$;

ues compared to the spherical particles. It can be concluded that the platelets contribute significantly to the stiffness of the material and are close to the reference material. Finally, the reduction in flexural modulus does not correspond to the reduction in fiber content, suggesting that the particles contribute to the stiffness of the composite.

4 Conclusion and Outlook

This study investigates the influence of particles of different sizes and shapes in endless glass fiber reinforced polymers and analyzes the effect on material composition, short beam strength and flexural modulus. It was found that the particle type significantly influences the material composition compared to the reference material. The fiber content decreases significantly with particle volume and the resin content increases in all samples. The short beam tests showed that the filler weaken the fiber and filler to resin adhesion. As the filler content increases and the resin content decreases, the short beam strength decreases significantly and the samples fail due to tensile stress. The flexural modulus also decreases as the fiber content is reduced, although the particles have a strengthening effect. The platelets act as crack initiators at only a few percent by volume, which significantly reduces the short beam strength. On the other hand, the platelets have a reinforcing effect on the flexural modulus compared to the spherical particles, almost reaching reference values. This study shows only one aspect of material behavior under mechanical loading. Another aspect, the tensile strength of particle filled fiber-reinforced composites, is considered in the study of Ganslmaier J. [17]. For future researches more particle sizes for the platelet shaped particles should be investigated. In addition, the electromagnetic absorption and the trade-off to the reduced mechanical strength of the materials must also be analyzed, strongly depending on particle size and shape.

References

1. Jölly, I., Schlögl, S., Wolfahrt, M., Pinter, G., Fleischmann, M., Kern, W.: Chemical functionalization of composite surfaces for improved structural bonded repairs. Compos. B Eng. **69**, 296–303 (2015). https://doi.org/10.1016/j.compositesb.2014.10.020
2. Liu, Y., Shigemoto, Y., Hanada, T., Miyamae, T., Kawasaki, K., Horiuchi, S.: Role of chemical functionality in the adhesion of aluminum and isotactic polypropylene. ACS Appl. Mater. Interfaces **13**(9), 11497–11506 (2021). https://doi.org/10.1021/acsami.0c22988
3. Arikan, E., Holtmannspötter, J., Zimmer, F., Hofmann, T., Gudladt, H.J.: The role of chemical surface modification for structural adhesive bonding on polymers - Washability of chemical functionalization without reducing adhesion. Int. J. Adhes. Adhes. **95**, 102409 (2019). https://doi.org/10.1016/j.ijadhadh.2019.102409
4. Li, H., et al.: A review of carbon fiber surface modification methods for tailor-made bond behavior with cementitious matrices. Prog. Mater. Sci. **132**, 101040 (2023). https://doi.org/10.1016/j.pmatsci.2022.101040
5. Xie, X., Yang, D.: Multi-functionalization of Al2O3 nanoparticles for enhancing thermal conductivity of epoxy natural rubber composites. Appl. Surface Sci. **602**, 154335 (2022). https://doi.org/10.1016/j.apsusc.2022.154335
6. Chandrashekar, A., Hegde, M., Krishna, S., Ayippadath Gopi, J., Kotresh, T., Prabhu, T.N.: Non-covalent surface functionalization of nanofillers towards the enhancement of thermal conductivity of polymer nanocomposites: a mini review. Eur. Polym. J. **198**, 112379 (2023). https://doi.org/10.1016/j.eurpolymj.2023.112379
7. Rudko, G., Kovalchuk, A., Fediv, V., Chen, W.M., Buyanova, I.A.: Enhancement of polymer endurance to UV light by incorporation of semiconductor nanoparticles. Nanoscale Res. Lett. **10**(1), 81 (2015). https://doi.org/10.1186/s11671-015-0787-5

8. Chikyu, N., Nakano, T., Kletetschka, G., Inoue, Y.: Excellent electromagnetic interference shielding characteristics of a unidirectionally oriented thin multiwalled carbon nanotube/polyethylene film. Mater. Des. **195**, 108918 (2020). https://doi.org/10.1016/j.matdes.2020.108918
9. Shukla, V.: Review of electromagnetic interference shielding materials fabricated by iron ingredients. Nanoscale Adv. **1**(5), 1640–1671 (2019). https://doi.org/10.1039/C9NA00108E
10. Timonen, J.V., et al.: Magnetic nanocomposites at microwave frequencies. In: Bârsan, V., Aldea, A. (eds.) Trends in Nanophysics, pp. 257–285. Springer, Heidelberg (2010). https://doi.org/10.1007/978-3-642-12070-1_11
11. Fu, S.Y., Feng, X.Q., Lauke, B., Mai, Y.W.: Effects of particle size, particle/matrix interface adhesion and particle loading on mechanical properties of particulate-polymer composites. Compos. B Eng. **39**(6), 933–961 (2008). https://doi.org/10.1016/j.compositesb.2008.01.002
12. Bard, S.: Kohlenstofffaser-Epoxydharz-Verbunde mit erhöhter Wärmeleitfähigkeit. Ph.D. thesis, Universität Bayreuth, Bayreuth (2021)
13. Rajaei, M., Wang, D.Y., Bhattacharyya, D.: Combined effects of ammonium polyphosphate and talc on the fire and mechanical properties of epoxy/glass fabric composites. Compos. B Eng. **113**, 381–390 (2017). https://doi.org/10.1016/j.compositesb.2017.01.039
14. Nordlund, M., Fernberg, S., Lundström, T.: Particle deposition mechanisms during processing of advanced composite materials. Compos. A Appl. Sci. Manuf. **38**(10), 2182–2193 (2007). https://doi.org/10.1016/j.compositesa.2007.06.009
15. Dosoudil, R., Franek, J., Slama, J., Usakova, M., Gruskova, A.: Electromagnetic wave absorption performances of metal alloy/spinel ferrite/polymer composites. IEEE Trans. Magn. **48**(4), 1524–1527 (2012). https://doi.org/10.1109/TMAG.2011.2172779
16. Hexcel Corporation: Product Data HexFlow® RTM 6 (2016). https://www.hexcel.com/user_area/content_media/raw/RTM6_RTM62_HexFlow_DataSheet.pdf
17. Ganslmaier, J., Santl, D., Buchenau, E., Bleckmann, M., Höfer, P.: Tensile behavior and failure mechanisms of carbonyl iron filled endless glass fiber-reinforced epoxy. In: Proceedings of the Munich Symposium on Lightweight Design 2024. Springer (2024). https://doi.org/10.1007/978-3-032-01360-6_3

Impact of a Reduced Holistic Process in the Concept Development of Giga-Casting Vehicle Components Regarding Weight Reduction Benefits

Florian Volk, Tobias Blichmann , Paul Zauner, Sebastian Schechner, and Patrick Schuster(✉)

Munich University of Applied Sciences, 80335 Munich, Germany
p.schuster@hm.edu

Abstract. This study emerged from a student project conducted as part of the Computational Engineering Master's Program at Munich University of Applied Sciences. The original task involved developing a giga-casting vehicle front-end based on the 2014 Honda Accord. The team's effort raised a question: What is the impact of different concept development processes on giga-casting vehicle components, particularly in achieving weight reduction benefits? This paper compares two distinct design approaches: a conventional method, which involves re-designing an existing component followed by parameter optimization, and an approach that begins with topology optimization, followed by the adaptation of the design to meet predefined parameters as well as a final design optimization (reduced holistic process). The project also evaluates the cost and environmental impact of the weight reduction within an innovative development environment. The final comparison assesses the weight difference of the developed concepts including impact on costs and the life cycle assessment. The result shows that in the innovative student project environment the consideration of a reduced holistic approach in the early stages of product development can significantly reduce component weight, costs, and environmental effects, compared to conventional design approaches.

Keywords: Giga-Casting · Reduced Holistic Process · Weight Reduction Impact · Concept Development · Cost Analysis · Life Cycle Assessment · Innovation

1 Project Set Up and Environment

1.1 Requirements and Set Up

As part of the Computational Engineering Master's Program at Munich University of Applied Sciences, eleven students were tasked with designing a giga-casting front-end for the 2014 Honda Accord (reference model in sheet metal design). The project included many requirements outside the development process itself, such as project management,

F. Endress et al. (Eds.): MSLD 2024, *Proceedings of the Munich Symposium on Lightweight Design 2024*, pp. 68–78, 2025.
https://doi.org/10.1007/978-3-032-01360-6_7

research, and more, to develop a component that is at least on par with the reference model. A detailed specification sheet outlined these requirements [1]. With no distributed budget, the project adhered to the courses time specification, which is based on the European Credit Transfer and Accumulation System (ECTS) [2] and translates into 1,600 student-hours over four months. The team exceeded this time allocation out of personal initiative, completing the project with three finished concepts, two of which followed entirely different design approaches (Sect. 2.3).

The reference model, sourced from the National Highway Traffic Safety Administration (NHTSA) [3], consisted of LS-Dyna crash simulation files containing the complete vehicle mesh structure of the Honda Accord. The weight and material data were included and derived from this dataset as well. To streamline the workflow, the nonlinear models were transferred into simplified linear models. The scope was narrowed to the front-end, simplifying data processing while satisfying project requirements.

A robust foundation for comparing the developed concepts was established by analyzing and evaluating the reference model. Simulations were conducted using educated estimates for the wheel forces. These calculated values did not need to be extremely accurate as their main use case was the comparison of the concepts with each other, and with the reference model. Scenarios such as cornering, emergency braking, and edge climb were considered, although dynamic and crash loadings were excluded per project requirements. Utilization factors, critical for comparing mechanical properties across concepts were calculated in accordance with the FKM-guideline [4]. This step was essential, given the absence of original development data from Honda.

1.2 Motivation

The industry's increasing adoption of giga-casting technologies inspired this project [5]. This was confirmed at one of the most renowned benchmarking conferences for car body development, the EuroCarBody, where original equipment manufacturers (OEM) that use big casting components were comparison winners based on the voting by international experts [6]. Initially intended as a design exercise, the team expanded its scope by investigating the comparative impacts of distinct design approaches. This was feasible due to the team's flexibility in working within a project environment, where established workflows or processes inexistant.

Transitioning established production systems to this technology can involve substantial changes and costs. By examining the component weight impact of differently designed concepts, it is to be determined whether these investments yield significant benefits.

Consequently, a research question evolved from the student project which initially aimed to develop a giga-casting front end with similar characteristics regarding weight, costs, environmental impact, and mechanical properties as the sheet metal design reference model. This question is: How does the concept development process influence the weight reduction, cost-efficiency, and sustainability of giga-casting vehicle components? Addressing this question required a deeper exploration of the design methodologies, peaking in the evaluation and comparison of the two approaches presented in this study.

2 Concept Development of Vehicle Components

2.1 Classic Design Approach

To achieve the goals of lightweight, cost, and environmental efficiency, it is necessary to consider and apply the relevant input data from different areas, such as production or design requirements, during product development.

Currently, the so-called "classic design process" is still often used in automotive practice. In this process (Fig. 1), separate analysis models are created for the individual requirement areas based on given CAD structures, each of which is loaded with independent input data [7]. This requires a sequential process, where the optimization only takes place within the considered area and in relation to its requirements [8]. The individual models cannot be directly linked to each other, so that the coordination between the individual domains is more complex. The individual models are harmonized in an additional model, which increases the computational effort and makes it more difficult to observe the necessary compromise of the combined models. This can lead to a suboptimal solution that does not fully exploit the potential of lightweight design [7].

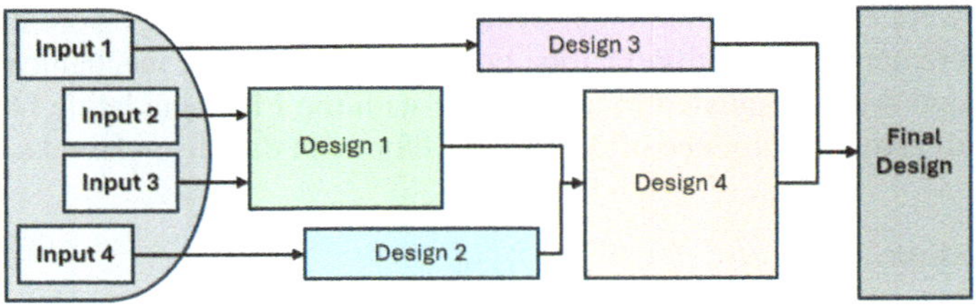

Fig. 1. Block diagram of the Classic Design Approach based on [7]

2.2 Holistic Optimization Approach

The approach is to create interdepartmental optimization models. These models combine the different requirements and input data to thus strengthen the realization of possible synergy effects. This approach is called the "holistic optimization approach" and consists of two phases (Fig. 2).

In contrast to the classic approach, there is initially no basic design of the component, which is why a topology optimization is performed in the first phase to approximate such a design. As part of the topology optimization, the load paths of the applied forces are generated, considering additional requirements from other areas, such as a direction of draw during the manufacturing process. The load paths are then used to derive a geometric model for further study. Once the first phase is completed, a design is available, that can be further optimized in the second phase to meet the specified criteria. This optimization takes place across all domains and considers all requirements in one model. Various optimization methods can be used, such as sizing, topology optimization, and freeform shaping, which can be applied specifically to components of the generated design.

At the end of the second phase, a final design that meets the objectives of all domains by creating synergies between the different domains without neglecting or excluding any of them is available [7].

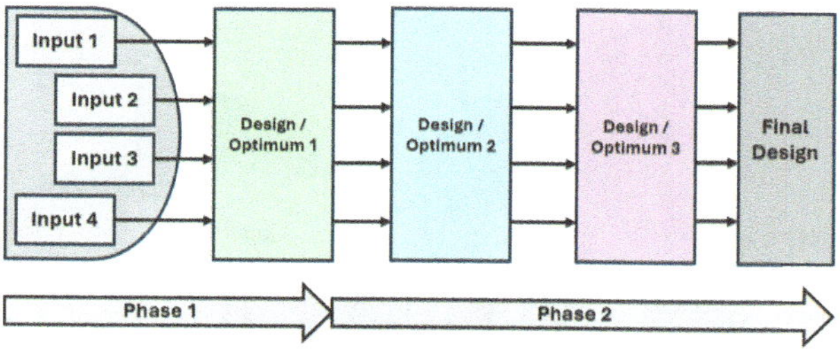

Fig. 2. Block diagram of the Holistic Design Approach based on [7]

2.3 Approaches Used in This Project

Two concepts based on the classic and holistic design approaches are developed for the design of a front-end giga-casting component.

The first concept uses the classic design approach. The starting point is the geometry of the reference front end, which is modeled into a demoldable flat structure that can accommodate the components of the base vehicle. Additional ribs are integrated to absorb the load in a low-speed crash and to improve stiffness. After verifying the required mechanical properties with a finite element (FE) model, the thickness of the structure is optimized to further reduce its weight.

From now on, it will be referred to a reduced holistic design approach rather than a holistic design approach. This change reflects the fact that in this project the team has not been able to fully implement the holistic approach described with the resources and capabilities available in the student environment.

The second concept is based on a reduced holistic design approach. In the first phase, topology optimization is performed. The design domain is derived from the original vehicle design. Input data includes the resulting forces for different load cases generated by the wheel forces acting on the chassis. Only the static load of a low-speed crash was considered to prevent significant deformations, which is deemed adequate for the initial phase of the project. As boundary conditions, the fixed support at the A-pillar and the sill as well as the connection points to the rest of the vehicle must be considered (Fig. 3). To reduce the computation time, an XZ-symmetry plane according to ISO 8855 is specified as a design constraint for the front end. As part of the concept development, specifications for the direction of draw are also defined to consider manufacturability. The results of the topology optimization are not used directly in the second phase of the development process but form the foundation for a surface-based design.

In the second phase, the thickness of the individual shell elements is used as a parameter for further weight reduction using shell thickness optimization. The goal of the optimization is to minimize the weight while maintaining stiffness, vibration, and

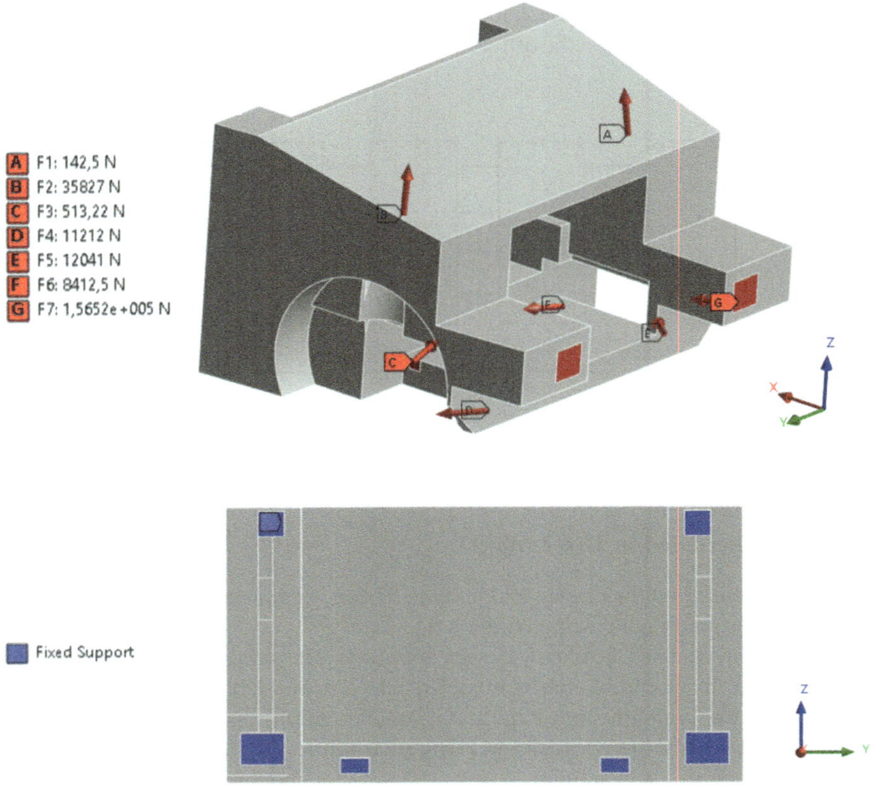

Fig. 3. Load cases (top) and boundary conditions (bottom) of the used design domain

material utilization constraints. First, a sensitivity analysis is performed to identify the parameters with the greatest influence on weight. This allows individual wall thicknesses to be combined into a common parameter, significantly reducing the number of variables and thus the computational time.

The defined parameter set, comprising a reduced number of variables, is subsequently transmitted to a solver for solution. Due to the project requirements, the integration of dedicated casting and explicit crash simulation as input variables within the reduced holistic development process was not mandatory. Figure 4 shows the results of the two design approaches. It should be noted that the strut braces are designed as separate cast components that are later joined to the front end. In this paper, the final front ends are examined as one-piece giga-casting components, independent of detailed casting parameters such as flow velocities, radii, and height differences.

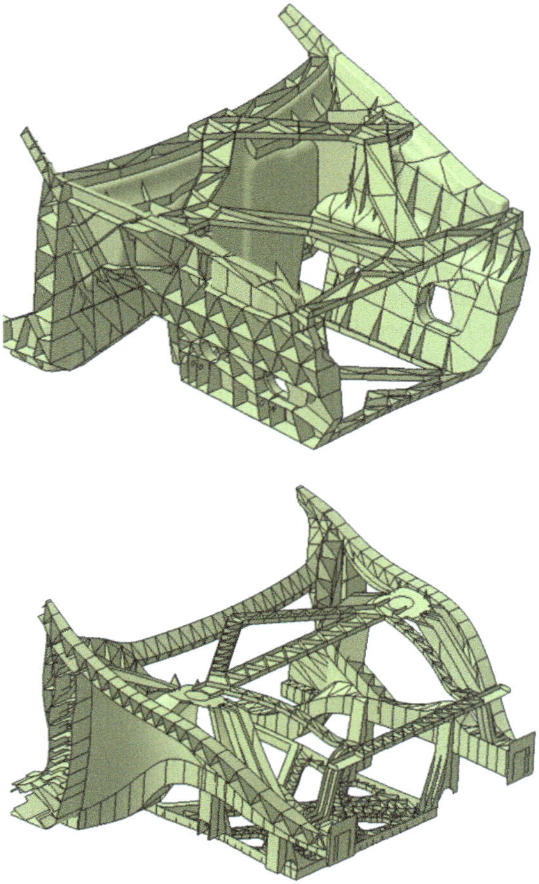

Fig. 4. Results of the classic (top) vs the reduced holistic (bottom) design approach

3 Impact and Results (Comparison)

3.1 Final Weight Reduction

A comparison of the results from the two processes reveals a significant weight advantage of the concept developed with the reduced holistic optimization approach. This is mainly due to the shell thickness optimization targeting weight reduction in low stressed areas. As a result, the weight was reduced by 33% compared to the classic approach. Comparing the giga-casting front end with the sheet metal design, a reduction in weight of 58% is achieved while the requirements related to mechanical properties are met and partially exceeded.

In the early stages of product development, decisions about cost and environmental factors are particularly relevant, as this is when the potential for influence is greatest. At the same time, it is often difficult to estimate the actual impact [9]. However, it is essential to consider cost and environmental aspects as they enable not only economic benefits, but also ecological progress and adaptation to regulatory requirements as well as changes in consumer behavior [10, 11]. In this context, the selection of the development process plays a critical role, as it significantly affects the weight of the component, particularly in the early stages, thus impacting both cost and environmental outcomes.

3.2 Cost Analysis

This study compares the costs of two die-cast aluminum concepts and the original model designed as a sheet metal design. To ensure a valid basis for comparison, the same methodology is used for both models. The material cost method according to VDI 2225 [12] is used as a basis. In this guideline, it is assumed that material costs represent a fixed proportion of the manufacturing costs. A material cost share of 30% to 45% was derived from literature as well as technical discussions [13, 14] and considered.

In addition, allowances for scrap (10%) and material utilization (40%) are considered using material cost factors. In concept development, as in this study, the allowances are typically based on estimates from available literature or known values, which means that the analysis is often subject to considerable uncertainty [15].

The results of the study, which are shown in Fig. 5, suggest that the reduced holistic approach with manufacturing costs between 620 and 950 euros is the more cost-effective option. In contrast, the classic approach has manufacturing costs between 1100 and 1700 euros, which is significantly higher. The difference between the concepts is due to the different weights and the range within the concepts is due to the material cost share.

In terms of material costs, the reduced holistic approach has a cost saving of ~40% compared to the classic approach due to the lower weight resulting from the reduced holistic optimization process. In addition, the material costs of the reduced holistic approach are at the same level as the sheet metal structure [16, 17].

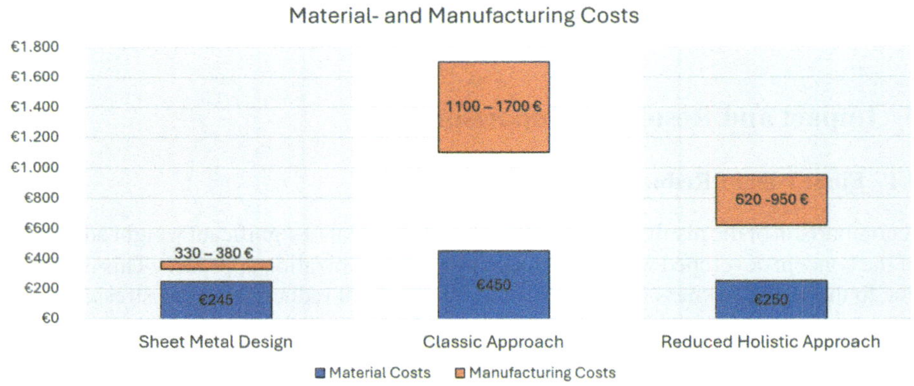

Fig. 5. Comparison of Material- and Manufacturing Costs

It should be noted that in addition to the production parameters mentioned above, such as scrap and material utilization, there are other factors that are subject to uncertainty.

In particular, investments in production equipment and savings through rationalization of possibly existing production processes, such as eliminating welds by switching to a casting component. This means that the informative value of the cost analysis is sufficient for initial comparisons, but not for reliable statements. The initial comparisons conducted lead to the conclusion that the reduced holistic approach can offer cost advantages based on weight reduction compared to the classic approach but is still less profitable compared to the initial model based on the data available for this study. However, a reduction in the variety of parts, the rationalization of production steps and personnel, as well as the required production area, cannot be sufficiently considered due to a lack of data.

Nevertheless, sources indicate that there is significant potential, so that in the future it may be possible to reduce manufacturing costs to a level comparable to the sheet metal structures. If it is possible to build production facilities from scratch, rather than converting existing facilities designed for sheet metal structures, the probability of becoming cheaper than sheet metal structures increases even further [18, 19].

3.3 Environmental Analysis

A life cycle assessment (LCA) based on DIN EN ISO 14040/14044 [20, 21] with the "cradle to gate" approach is used. The analysis refers to Germany in the year 2024. Due to the limited availability of specific process data, general assumptions are made, particularly with regard to the utilization rate and production waste. The database ProBas [22] and the analysis software OpenLCA [23] are used. As part of the environmental analysis, CO_2 emissions are evaluated as a key environmental impact. The evaluation is performed using the Midpoint ReCiPe2016 method [24].

The results suggest that the classic approach is associated with higher overall emissions due to higher material consumption. The reduced holistic approach has advantages due to the reduced component weight, which leads to lower CO_2 emissions, especially in the extraction of raw materials. The evaluation is carried out using different proportions of secondary aluminum. A maximum of 40% is recommended for 2024, as otherwise the mechanical properties of the cast alloy are no longer guaranteed [26]. Nevertheless, the results are also subject to uncertainties, as specific process data for giga-casting is not available. For example, this study assumed a CO_2 equivalent of 4 kg per kg aluminum, which is the average in Germany in 2024, while the global average is 14.8 kg. [25, 26].

On-site foundry and logistics expenses are also key factors in determining the level of environmental impact. In terms of environmental impact, the classic approach in this project has no advantages over the sheet metal structure. The reduced holistic approach proves to be more advantageous than the initial model at a certain level of secondary aluminum content (Fig. 6). Nevertheless, it can be predicted that the CO_2 footprint of cast aluminum alloys will be significantly reduced in the coming years and that the quality of secondary aluminum will also be optimized [27, 28]. This could lead to new concepts that outperform the original model in terms of their environmental footprint.

In the discussion it is pointed out that the significance of the environmental analysis in this project is limited. Manufacturers who have detailed process information will be able to make more accurate assessments and thus reliably determine the advantages and disadvantages of the approaches with fewer deviations.

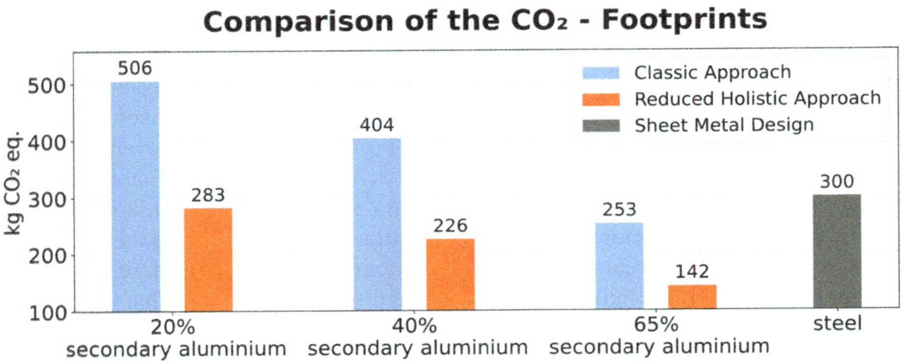

Fig. 6. CO_2 footprint with different amounts of secondary aluminum.

4 Conclusion

In summary, the reduced holistic development approach can provide substantial advantages, particularly in terms of flexibility and efficiency. This approach enhances adaptability to varying conditions while allowing a focus on critical elements, decisive within innovative project environments and early development stages. The upstream topology optimization offers the opportunity to improve front-end properties, crucial for the automotive industry facing increasing innovation pressures [29]. Additionally, it allows for the consideration of numerous variants during the concept phase, helping to find balanced solutions to diverse interdisciplinary demands. However, many parameters, influenced by the number of input parameters, boundary conditions, and concept variants, can slow and complicate the process due to significant computational demands.

In contrast, the conventional development process may become more time-intensive when attempting to achieve outcomes like those possible with a reduced holistic approach. Repetition or even serial use of the processes might change the outcome, then however, the project environment is no longer as innovative and therefore not comparable. The classic approach heavily relies on empirical data and requires the integration of all input data from the beginning, which can complicate making changes to the foundational structure, especially in complex and innovative areas such as giga-casting of large vehicle components.

The advantages in the process are also reflected in the significantly lower weight of the second concept. By using the reduced holistic development approach, the weight could be reduced by 33% compared to the first concept with the classic development process.

The casting concepts developed on the basis of the available data do not show any advantages in terms of manufacturing costs. However, the reduced holistic approach resulted in significantly lower costs than the classic approach. The potential for cost reduction for the giga-casting concepts can be considered significant as the technology is still relatively new.

In terms of environmental impact, the reduced holistic approach performs better than the classic approach. In addition, the reduced holistic approach performs notably better than the initial model when 20% secondary aluminum is used, while the classic approach breaks even at 60% secondary aluminum.

It is predicted that significant improvements in both manufacturing costs and environmental impact will be achieved in the coming years through process improvements, rationalization, and the use of environmentally friendly aluminum alloys [27, 28]. For users who have the option of building a greenfield production facility in combination with using a reduced or even holistic process, giga-casting is a promising option. The reason is that in this environment the structure of the production lines and logistics can be designed almost freely, while being optimally adapted. This is directly reflected in the manufacturing costs and environmental impact.

By considering a reduced holistic process in the early stages of giga-casting vehicle product development in an innovative project environment, component weight can be significantly reduced, compared to conventional design approaches. This positively influences costs and environmental impact.

Acknowledgements. This research was based on the work of the project team "Konzeptentwicklung mechanischer Strukturen" in the summer semester 2024 at Munich University of Applied Sciences, supervised by Prof. Dr.-Ing. Klemens Rother.

References

1. Rother, K.: Munich University of Applied Sciences, Lecture Document TBM Master SoSe24: Projektbeschreibung im Modul Konzeptentwicklung mechanischer Strukturen. Not published (2024)
2. Atack, P.: Study.eu, Online Article: What is the European Credit Transfer System (ECTS)? (2024). https://www.study.eu/article/what-is-the-ects-european-credit-transfer-and-accumulation-system. Accessed 15 Dec 2024
3. National Highway Traffic Safety Administration, Website: Crash Simulation Vehicle Modells. https://www.nhtsa.gov/crash-simulation-vehicle-models. Accessed 01 July 2024
4. Rennert, R.: Rechnerischer Festigkeitsnachweis für Maschinenbauteile aus Stahl, Eisenguss- und Aluminiumwerkstoffen - 7. Auflage, VDMA Verlag, Frankfurt am Main (2020)
5. Kopp, T.: BMW Group, Seminar at BMW Factory Landshut: Gießtechniken und Metallguss. Landshut (2024)
6. Automotive Circle, Conference Website: EuroCarBody 2024 - 26th Global Car Benchmarking Conference. https://www.automotive-circle.com/events/eurocarbody-2024/. Accessed 04 Dec 2024
7. Triller, J., Lopez, M.L.: Multidisciplinary optimization of automotive mega-castings merging classical structural optimization with response-surface-based optimization enhanced by machine learning. Sci. Rep. **13**(1), Article No. 21678 (2023)
8. Schönmann, S.O.: Produktentwicklung in der Automobilindustrie – Managementkonzepte vor dem Hintergrund gewandelter Herausforderungen. Springer, Wiesbaden (2012)
9. Ehrlenspiel, K., Kiewert, A., Lindemann, U., Mörtl, M.: Kostengünstig Entwickeln und Konstruieren - Kostenmanagement bei der integrierten Produktentwicklung. Springer, Berlin (2020)
10. Umwelt Bundesamt: Neue Ökodesign-Verordnung für nachhaltige Produkte tritt in Kraft (2017). https://www.umweltbundesamt.de/themen/neue-oekodesign-verordnung-fuer-nachhaltige. Accessed 18 Dec 2024

11. PWC: Klimafreundlicher und nachhaltiger: Die Deutschen passen Einkauf und Konsum an (2024). https://www.pwc.de/de/pressemitteilungen/2024/klimafreundlicher-und-nachhaltiger-die-deutschen-passen-einkauf-und-konsum-an.html. Access 18 Dec 2024
12. VDI 2225 Blatt 2, VDI Guideline: Konstruktionsmethodik - Technisch-wirtschaftliches Konstruieren. Beuth Verlag, Berlin (1998)
13. Fritsch, T., Sammer, K., Kopp, T.: Alles Giga oder Mega? Es gibt kein „richtig" oder „falsch"! (2024). https://www.foundry-planet.com/d/everything-giga-or-mega-there-is-no-right-or-wrong/. Accessed 19 Dec 2024
14. Roos, H.J., Lagler, M., Quintana, L.: Company document: Die Zukunft von Strukturbauteilen für Druckgusslösungen. https://dam.buhlergroup.com/asset/86bd505f19e643b2b3ce79957169681c. Accessed 06 July 2024
15. Herrmann, C., Pries, H., Hartmann, G.: Energie- und ressourceneffiziente Produktion von Aluminiumdruckguss. Springer, Heidelberg (2013)
16. Worldautosteel, Article: ULSAB AVC - Advanced Vehicle Concepts - Cost Models (2015). https://www.worldautosteel.org/projects/cost-models/. Accessed 06 July 2024
17. WirtschaftsVereinigung Metalle. e.V., E-Mail-Communication: Metallpreisübersicht (2024)
18. Li, T., Song, J., Zhang, A., et al.: Progress and prospects in Mg-alloy super-sized high pressure die casting for automotive structural components, Chongqing University, Chongqing (2023)
19. Hartlieb, A., Hartlieb, M.: The Impact of Giga-Castings on Car Manufacturing and Aluminum Content (2023). https://www.lightmetalage.com/news/industry-news/automotive/the-impact-of-giga-castings-on-car-manufacturing-and-aluminum-content/. Accessed 08 July 2024
20. DIN EN ISO 14040, DIN standard: Umweltmanagement - Ökobilanz: Grundsätze und Rahmenbedingungen. Beuth Verlag GmbH, Berlin (2021)
21. DIN EN ISO 14044, DIN standard: Umweltmanagement - Ökobilanz: Anforderungen und Anleitungen. Beuth Verlag, Berlin (2021)
22. Umweltbundesamt, Online database: Probas. https://www.probas.umweltbundesamt.de/. Accessed 21 May 2024
23. GreenDelta GmbH, Software: openLCA (2024). https://www.openlca.org/download/. Accessed 25 May 2024
24. Huijbregts, M.A.J., Steinmann, Z.J.N., Elshout, P.M.F., et al.: ReCiPe2016: a harmonised life cycle impact assessment method at midpoint and endpoint level. Springer, Heidelberg (2016)
25. Kneer, S.: E-Mail-Communication: Albert Handtmann Metallgusswerk GmbH & Co. KG, Informationen Giga-Casting. München (2024)
26. Lagler, M.: E-Mail-Communication: Bühler AG, Informationen Giga-Casting. München (2024)
27. Albert Handtmann Metallgusswerk GmbH & Co. KG, Company document: Klimastrategie 2039, Leichtmetallguss (2024)
28. Prockl, S.: Company document: Bühler AG, Sustainable megacasting production for the body-in-white of tomorrow (2024)
29. Duddeck, F.: Multidisciplinary optimization of car bodies. Struct. Multidisc. Optim. **35**(4), 375–389 (2008)

Author Index

F. Endress et al. (Eds.): MSLD 2024, *Proceedings of the Munich Symposium on Lightweight Design 2024*, p. 79, 2025.
https://doi.org/10.1007/978-3-032-01360-6

The manufacturer's authorised representative in the EU is Springer Nature Customer Service Centre GmbH, Europaplatz 3, 69115 Heidelberg, Germany. If you have any concerns regarding our products, please contact ProductSafety@springernature.com

Printed and bound by CPI Group (UK) Ltd, Croydon, CR0 4YY
30/09/2026
02225575-0002